ISLANDGODS

EXPLORING THE WORLD'S MOST EXOTIC ISLANDS

RICHARD BANGS

PHOTOGRAPHY BY PAMELA ROBERSON

TAYLOR PUBLISHING COMPANY • DALLAS, TEXAS

To the memories of Vera "Bobbie" Roberson,
who was born on an island, and to Bob Yost,
ambassador to an island, and a friend

Also by Richard Bangs
Rivergods (with Christian Kallen)
Islands of Fire, Islands of Spice (with Christian Kallen)
Riding the Dragon's Back (with Christian Kallen)
Paths Less Travelled (with Christian Kallen, editors)
Whitewater Adventure
Adventure Vacations (editor)

Published by Taylor Publishing Company
1550 West Mockingbird Lane
Dallas, Texas 75235

Calligraphy and Design by Walter Gray Lamb

Maps by Maryland CartoGraphics Inc.

Library of Congress Cataloging-in-Publication Data

Bangs, Richard, 1950–
Islandgods : exploring the world's most exotic islands / Richard Bangs and Pamela Roberson.
p. cm.
Includes index.
ISBN 0-87833-742-3 : $39.95
1. Islands. I. Roberson, Pamela. II. Title. III. Title: Island gods.
G500.B36 1991
909′.094′2—dc20
91-17068
CIP

Printed in the United States of America

10 9 8 7 6 5 4 3 2 1

Acknowledgments

No book is an island. Writers and photographers work alone, inspiration tumbling forth from some magical inner void that is mysterious even to them. But the final product is a collaborative effort. It takes an archipelago of people and organizations to produce a work such as this, and we would like to acknowledge some of those whose generous help, guidance, and support created the collection of paradises you now hold in your hands.

First and foremost we want to acknowledge Michael Fagin, the president of the St. Louis-based International Group. Mike's enthusiasm for this book was unbounded, and he spent hours, no, days, perhaps weeks, spinning through his very fat Rolodex culling the contacts needed to make this ambitious project work. Without our friend Mike there would have been no book.

Many assisted us with the specialized and quality gear needed to explore these islands. They include: Lee Turlington and the company he's helped guide to great success, The North Face, manufacturers of the finest camping gear and clothing in the world; Christa Kaiser, president, International Divers, Inc., who supplied the diving gear; Alan Berry of Bucci Sunglasses; Minolta cameras and lenses; Kodak film; Insul-Mats from Gymwell Corporation, who gave us sleeping comfort in some rough places; the finest adventure footwear ever, Teva Sport Sandals, and Mark Thatcher, their inventor; and Eagle Creek luggage.

Getting to and around each island was a monumental task, and many helped to ease the passage. These include:

Bahrain: Mohamed Hakki at the National Press Building in Washington, D.C.; Sheikh Rashid Bin Khalifa Al-Khalifa; the Minister of Information, Tariq A. Almoayed; Mirza Al Nasheet and Adel Abdulla Ahmed of Tourism Projects Company; Al Sayed Al-Bably of the Prime Minister's Office; TWA and Cathay Pacific Airlines.

Dominica: Prime Minister Eugenia Charles; Minister of Agriculture and Tourism Charles Maynard; W. Ken Alleyne, General Manager of the Dominica National Development Corporation; Marie-Jose Edwards, Director of Tourism; Gerry Aird, Chairman of Tourism; Derek and Ginnette Perryman of Dive Dominica; the Fort Young Hotel; the Papillote Wilderness Retreat; Christian Karan of the Coconut Beach Hotel; American Airlines; Leeward Islands Air Transport; British West Indies Airways; Ken George-Dill of Ken's Hinterland Adventure Tours; the famous mountain climber, Carnette Pemberton; Parliament member Ann Timothy; our driver and patient model Clem Johnson; and Lennox Honychurch.

Easter Island: Mike Gallegos of Latour of Santiago; Diana Samper of Ladeco Airlines; UTA French Airlines; Marilyn Pugh of Lan Chile Airlines; the Tahiti Tourist Board; Victoriano Girale of Kia-Koa Tours; Hotel Iorana; guide Hugo Hey; Edmundo Edwards; and Jules Wong.

Honshu: Ray Benton of Travent International of Waterbury Center, Vermont; old friend and survivor Rick Laylin of United Airlines; Tada Yasue of Skyland Holidays; Patrick Delaney, Richard Weiss and Yoshi Makiuchi of Travent; and West J.R., Sobek's travel partner in Japan.

Lombok: Dr. Halim Indrakusuma, Director of Pacto Tours; Garuda Indonesian Airways; Peter Arya of the Legian Beach Hotel in Bali; the Senggigi Beach Hotel; Chris Dawson, who shared the secret of this special island with us; Chris Drew of Sports Marketing Enterprises.

Madagascar: Roger Rakotomalala, of Lemur Tours in San Francisco, and Michele Randrianaridera, the general manager of Lemur Tours in Madagascar; Inder Sethi of Air Mauritius; Air Madagascar; Vijaye Haulder, of the Mauritius Government Tourist Office; the Royal Palm Hotel of Mauritius; Dr. Alison Jolly; the Hilton Hotel and the Hotel Colbert in Antananarivo; Les Cocotiers Hotel in Nosy Be; the Berenty Reserve; Hotel Buffet de la Gare in Perinet; Chez Claudine, with its uncommon service and fare; and Susan Dworski, whose enthusiasm for the Great Red Island was contagious.

Newfoundland: Richard Deacon of Sobek-Canada; Mark Dykeman and Jim Price of Eastern Edge Outfitters; Margie Price for her incredible cookies; Frank Leggett and Air Canada; Susan Sherk of the Economic Recovery Commission; the Compton House in St. John's; the Victorian Manor of Trout River; Ocean View Motel of Rocky Harbour; and the Driftwood Inn of Deer Lake.

Sri Lanka: Air Lanka; Tilak Fernando of the Ceylon Tourist Board; the Oberoi Hotel in Columbo; the Moonlight Beach Hotel in Nilaveli; the Citadel in Kandy; and Korean Air.

Tasmania: Al Baker and Richard Corbet of World Expeditions; Ed Hamilton of the finest island airline in the world, QANTAS Airways; the Sheraton of Hobart; superguides John Wier and Andrea McQuitty; Kevin Arnold and David Walker of Tourism Tasmania; Dr. Bob Brown of the Tasmanian Parliament; Robert Heazlewood of Eyelevel, who arranged a helicopter for an aerial view; Bob Burton of the Wilderness Society; the New South Wales Tourist Commission; the Intercontinental Hotel in Sydney; Goronwy Price of Australian Direct Travel; and Lord Sir Richard Taylor for his droll humor.

Trobriands: Jan and Peter Barter and Evelina Stanley-Knossen of Melanesian Tourist Services and the Melanesian Discoverer; Kerry Byrd and Jane Allen of Air Niugini; QANTAS Airways and Ed Hamilton, Sobek's district sales manager; Talair; the Islander Hotel in Port Moresby; the Kirawina Lodge in Losuia; and my hero, Bill Rudd.

Others who assisted in various ways: lounge singer and good friend Steve Marks of ICM; Hap Klopp of HK Consulting; Jens Ingolfsson of Oden Air in Iceland; Kitnasamy Alwar of Mauritius Tourist Office; and Dennis Graver of NAUI.

Diana St. James, who helped with endless arrangements and manuscript versions; Melanie Tan, who did a herculean job of keeping the accounts in order; Tricia Newkirk of Sobek, who helped with all the air arrangements; Dr. George Fuller, who helped whenever there was a computer snafu; John Yost, intrepid partner in Sobek; Russ Daggatt and Sharon Ahern, who covered for us while we were away; Johannes Tan, who designed the proposal for this book; Mary Crowley of Ocean Voyages; Terry and Linda Thompson, who loaned Pelican cases, and also birdsat; Marti Morec, whose imaginative research provided the fascinating marginalia; the El Rancho Best Western in Milbrae, California; Bill Graves of *National Geographic*; the World Wildlife Fund; Peter Guber of Columbia Pictures; and the rest of the Sobek staff.

And Kathy and Kelly Hamlet, who took care of Bali and Borneo and Xingu while we were off in faraway lands. Frank Roberson and Dr. and Mrs. L. C. Bangs, who bear the ultimate credit for this and other creations.

Finally, Jim Donovan of Taylor Publishing, who believed in this project when it was just a list of islands on a piece of paper.

Contents

EXOTIC ISLANDS OF THE WORLD

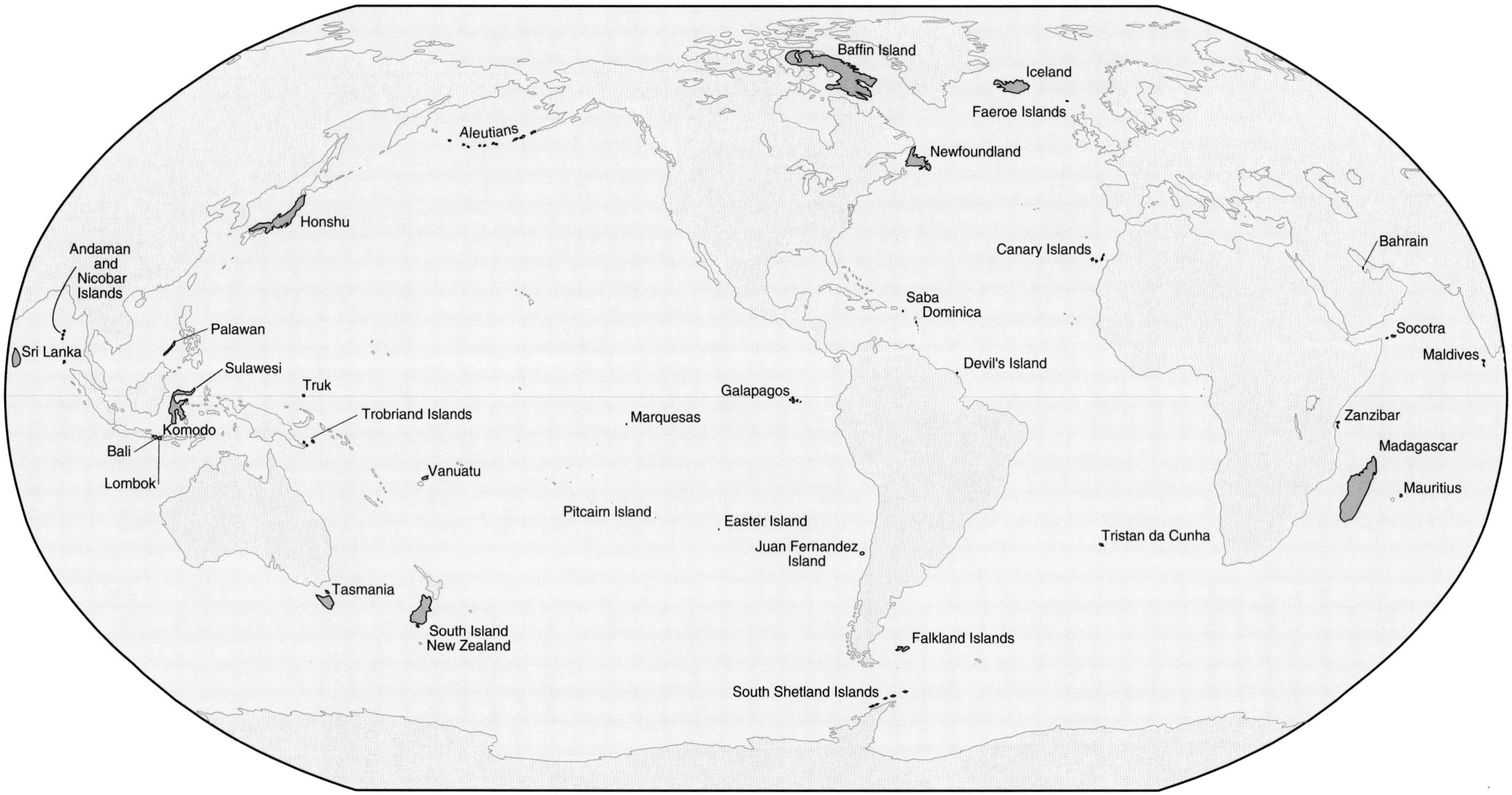

Baffin Island
Iceland
Faeroe Islands
Aleutians
Newfoundland
Honshu
Andaman and Nicobar Islands
Canary Islands
Bahrain
Saba
Dominica
Palawan
Sri Lanka
Socotra
Maldives
Sulawesi
Devil's Island
Truk
Galapagos
Trobriand Islands
Marquesas
Komodo
Bali
Zanzibar
Madagascar
Lombok
Vanuatu
Mauritius
Pitcairn Island
Easter Island
Juan Fernandez Island
Tristan da Cunha
Tasmania
South Island New Zealand
Falkland Islands
South Shetland Islands

Introduction

This other Eden, demi-paradise,
This fortress built by Nature for herself
Against infection and the hand of war,
This happy breed of men, this little world,
This precious stone set in the silver sea,
Which serves it in the office of a wall,
Or as a moat defensive to a house,
Against the envy of less happier lands . . .
—WILLIAM SHAKESPEARE, *King Richard II*

Now will I believe that there are unicorns.
—WILLIAM SHAKESPEARE, *The Tempest*

When I was twelve I discovered maps. I would pore over the contour lines and meridians, and conjure up the mysterious unknown places. But I was particularly drawn to the void spaces on the charts, where occasionally a dot, an island half stolen from the sea, half from my imagination, would call and grab me by the scruff of the neck, pulling me in.

"How could there be latitude and longitude to such a thing of dreams and fancy?" asked nineteenth-century travel writer Augusta de Wit when she visited the Dutch East Indies in the first decade of this century. "An attempt at determining the acreage of the rainbow, or the geological strata of Fata Morgana, would hardly have seemed less absurd. I would have none of such vain exactitude, but still chose to think of this place as situated in the same region as the Island of Avalon, the Land of the Lotus Eaters, palm-shaped Bohemia by the sea, and the Forest of Broceliand, Merlin's melodious grave. And it seemed to me that the very seas which girt those magic shores, keeping their golden sands undefiled from the gross clay of the outer world must be unlike all other water—tranquil ever, crystalline, with a seven-tinted glow of strange sea flowers, and the flashing jewel-like fishes, gleaming from unsounded deeps."

Islands have always provoked the poetry within us, have always tempted us with visions of paradise on this rough earth, of exotic, sensuous retreats sans cares or confusion; only sands of indolence and hedonism. And islands have always been the abodes of gods, of icons and divine beings who shaped the landscape, and the personality of its peoples.

Islands, too, have been sanctuaries from the ravages of Man and his ambitious works. While the great rivers of the continents have been dammed and the cities smothered in their own pollution, the relative isolation and size of islands have often left them alone with sometimes singular ecologies, and a degree of environmental, cultural, and spiritual integrity.

It was this element of integrity that drew photographer Pamela Roberson and myself to the islands herein. Our plan was to circle the globe, in no particular pattern, with no set schedules, and survey the splendor of the planet through its islands, to pull these places into an orbit of personal compassion. It was a trip that took eleven years, and the time on any island was always far too short. Variety is the facile handmaiden of islands, and as such we chose the ten islands featured because of their geographic, biological, and cultural diversity, a sort of global necklace of varied habitats, peoples, and beliefs. We also selected them for an elusive quality we might as well call romance, the honey that pours over a piece of real estate and makes it glow under the sickle moon. Somehow these islands, part of the web that holds our lives together, caught us in their spell, and we surrendered.

Because romance holds hands with adventure, we wanted to reconnoiter these islands, not by tour bus or taxi, but in adventurous ways. And so we kayaked, sailed, dived, cycled, climbed, rafted, rode horseback, rode trains, trundled, and sometimes crawled through these landscapes. And, as these were lands with uncertain borders between the temporal and spiritual, we touched the cosmologies of Hindus, Moslems, Christians, Buddhists, animists, and atheists. We met adventurers, heroes, scholars, charlatans, and brutes. We came face to face with moose, camels, caribou, whales, lemurs, geckos, platypuses, crocodile handbags, and the Arabian oryx, the straight-horned antelope thought to be the legendary unicorn. We trekked through rain and snow, across arid wastes, through luxuriant rain forests, up fiery volcanos, beneath cool pines and sweaty palms. We lost luggage and tempers, ate too little and too much, fell ill, missed boats and planes, bruised skin and egos, and fractured several foreign languages. Along the way we rummaged through Nature's attic, where we discovered many wonderful things; plants that forgot their manners and behaved like trees, reefs that roared like lions, burning blue skies and kohl-rimmed eyes, filigree-fine music and perfumed air. But we also saw where things were going wrong: where nature coexisted uneasily with human poverty; where swaths of forest had been reduced to cinders; where people and land are overtaxed; where wildlife with saucer eyes sits on the edge of extinction.

Not so long ago the islands we visited held a fraction of the people as now, and twice as many animals. No engine roared louder than the wind, and at night the moon outshone all other lights. There were uncharted interiors, lost hills, and hidden valleys. These islands were antediluvian life rafts, living museums of the best the earth had to offer.

Now things are different. There is a mutiny of the bounty. The treasures of these islands are being drilled, mined, plowed, and plundered; the Ark is sailing through a whirlpool that threatens to spill its precious cargo. And if the boat goes over, what will be left is history, and morality gone amok. There are no lines in the sea, but we seem to have crossed one nonetheless, and are now trespassing, disregarding nature's well-marked signs at our own peril.

Islands may be the last chance to preserve the finest pieces of our miracle planet. And as such, this book is a feast of sea-girted wildness, a seductive voice designed to lure our senses into awareness, a rococo show of the dignity of nature, and finally a reason to save the unicorns.

Like a cool, feminine hand over a throbbing forehead, islands soothe the countenance of a mad and feverish world. They are necessary *elixir vitae,* sweet tonics to quaff at the end of a long day. And so within these pages we celebrate ten great islands of the seven seas, and the spirits and gods that steer their courses. In the back of the book we also include a raft of "lesser gods," using the same insufficient criteria as for the featured chapters. However, I must confess this may be the most incomplete book ever published, as there are as many islands as there are dreams. These, though, are the paradises of our hearts and we invite you to paddle there with us, to wallow in the swells, to sip a cool drink with a little umbrella stuck in the pineapple, and explore the acreage of a rainbow.

Richard Bangs

The Island of Laputa, May 1991

Dominica

The Edge of Eden

Keep a green tree in your heart, and perhaps a singing bird will come.
—Chinese proverb as quoted on a Caribbean Conservation Association poster

We have scotched the snake, not killed it.
—William Shakespeare

It crawls from the sea like a dusky, ridged, primordial creature, twisting upwards into the *chatannye* trees, a serpent entering the Garden of Eden, delivering some fatal gift. The local Caribs call it *L'Escalier Tete-Chien* ("The Snake Staircase" in French patois), and it snakes up the east coast of Dominica, the most mountainous of the anglophone Windward Islands in the West Indies, a place Columbus sailed by five hundred years ago, as most tourists do today. This natural lava-flow staircase twists up the southern end of the Carib Indian Territory, the last reserve of indigenous Caribbean people in the world. Though today most of the three thousand surviving Caribs have succumbed to Catholicism, a few cling to the ancient myths and lore of a time before skins of white and black invaded their secret gardens.

Euphrasen is a copper-skinned, middle-aged woman with high cheekbones, blue-black hair, and delicate Asian eyes. She tends the kiosk on the narrow road near the crest of L'Escalier Tete-Chien at the village of Sineku. Euphrasen told me that Ma Janey Valmon, the oldest living Carib (she guesses 107, but has no birth records), knew many things about the Snake Staircase. Soon after I arrived on Dominica, I saw Janey stooped in a doorway, puffing on a cigarette—her skin the color of yellow parchment, one eye lost to some long-forgotten ravage of age. A former midwife and witch doctor, she was still lucid and feisty, and when I asked directions to L'Escalier Tete-Chien she pointed a craggy finger to the south. From Ma Janey, Euphrasen had heard the story of the giant serpent that long ago emerged from the sea and crawled to the mountains, leaving in its tracks the grim staircase. The legend claimed that if one walked the staircase from the sea to its summit during the phase of the full moon, enlightenment would be achieved. Euphrasen had tried it several times, but had ended her climb at the kiosk, as across the road there was only thick jungle. No sign of a continuing staircase. Though she was a happy woman, she wasn't sure she had found the promised enlightenment. I thought perhaps there was something beyond the road, another step to the promised higher ground, and I hoped to explore it.

It was late afternoon on April 10 when I bought a Fanta orange soda from Euphrasen's kiosk, and the full moon was just peeking over the Atlantic. I was thirsty. The trek up the Snake Staircase was steep, though not that long, less than a mile. But the day was hot and humid, and I wasn't in peak condition. Still, I had made the pilgrimage, had touched

Opposite. Prince Rupert Bay. Because of its shallow surrounding reefs no major cruise ships dock off the island, and thus its beaches remain pristine, with very few tourists.

the sea water at the tenebrous base of L'Escalier Tete-Chien. I had watched the crashing surf for several minutes, the brilliant white foam slapping the burnished black basalt, the pale spray blowing about like smoke. Then I turned my eyes to the staircase, which seemed to move like a mirage, as though alive. Motion illusion, I knew—something I'd experienced many times while gazing at a rapid on a wild river, then turning eyes to shore, which would then undulate for long seconds. But here, as the rock beneath me seemed to billow, almost as though breathing, I found myself spreading my arms and legs for balance. Almost instantly the sensation subsided, and the terra was once again firma.

I turned and climbed the high steps towards the plateau. To me the rock I straddled looked like an ancient lava dyke, one that oozed from some caldera in the center of the 290-square-mile island. The softer sedimentary rock on either side of the dyke had eroded, leaving the elevated dark rock causeway. In cooling, the lava had cracked into natural steps that climbed from the sea, but its course petered out at the kiosk, as I had. I ordered another Fanta, feeling no more or less enlightened than before the trek, just thirstier.

That night, as the rain drummed on the roof of the Fort Young Hotel in Dominica's sleepy capital of Roseau, I spread a topographic map across the terrazzo floor and traced the contours with my fingers. The lava that created L'Escalier Tete-Chien could have come from almost anywhere. The almond-shaped island itself was created from epeirogenic uplifts and submarine eruptions some twenty-six million years ago, with the last volcanic explosion just over a century back, and rumblings as recent as 1975. It might have come from Morne Diablotin, at 4,747 feet the highest peak on this Caribbean isle. If the lava was relatively new, by geological standards, it likely flowed from the 16,000-acre

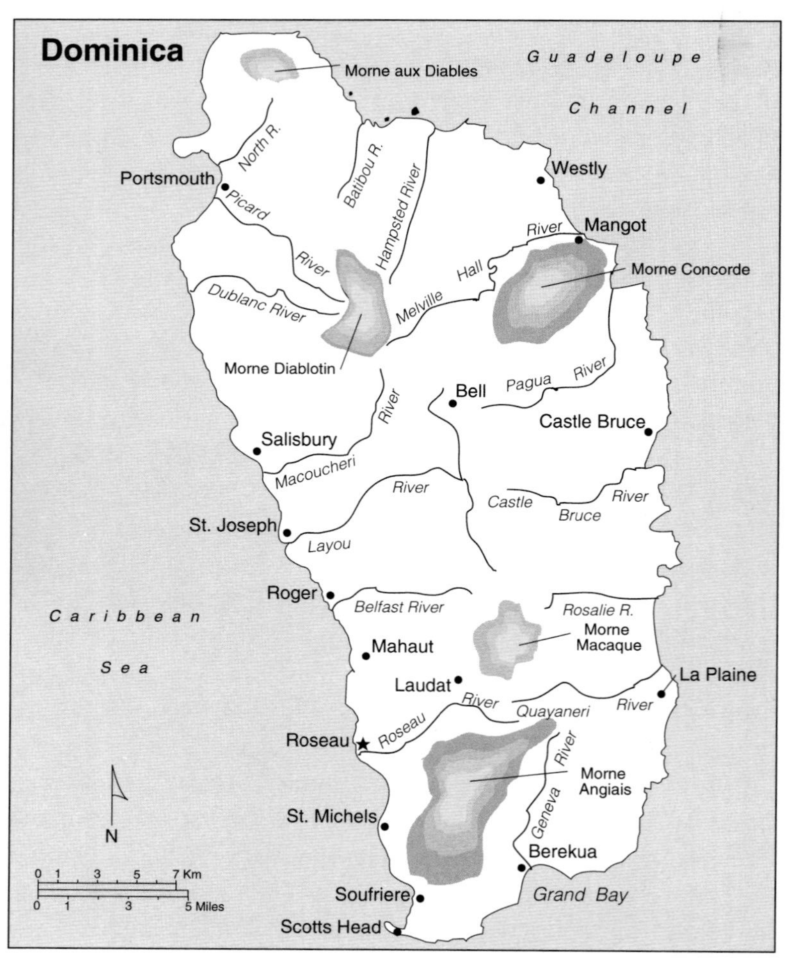

Below. The beaches of Dominica are mostly of black sand, the result of volcanic activity over the centuries. *Right.* The village of Soufriere on the southwestern coast of Dominica.

Morne Trois Pitons (Mountains of Three Peaks) National Park, the most volcanically active region on this real-life Skull Island. And if it flowed within the last thousand years, it may have come from the Grande Soufriere, or Boiling Lake—a Dante-esque cauldron, the largest of all the quiescent craters of the Caribs, and the second-largest thermally bubbling body of water in the world (the largest is in New Zealand). That would be my target.

The full moon was setting over the Caribbean, on the western side of the island, as we piled into our blue L300 Mitsubishi van at dawn and trundled six miles up the Roseau Valley to the village of Laudat, a collection of wide-planked pastel houses on stilts almost two thousand feet above the sea. Along for the trek was photographer Pamela Roberson and our guide, Clem Johnson, a twenty-seven-year-old Cambridge-educated Dominican who had returned to his island to seek his fortune. Clem had been ushering us around the island for several days, showing off its secrets and complaining about governmental policies that might compromise its future. He had hoped to become a nature photographer, but all his camera gear, $6,000 worth, had been stolen two weeks ago on the northeastern windward coast. Now he was a hired guide, a shepherd in the tourism business, a business he believed could discredit the integrity of the island. I liked Clem, and agreed with his assessments of the downside of tourism. Though he seemed to smoke a pack a day of the Dominican-made Hillsborough Special filter-tipped Virginia cigarettes, Clem was in good shape. This would be his sixth climb to the Boiling Lake.

Columbus. The island received its name from the Latin *dies dominica* (the Lord's day, or Sunday), the day on which Christopher Columbus sighted it in 1493.

After parking the van we stepped up to a wooden platform and followed its winding path alongside a tar-covered aqueduct pipe that looked like a giant black snake. The platform and pipe terminated at the diggings of a cement-lined water tank. Bulldozers, tractors, and other machinery of progress surrounded us. This was one of the sites of the Hydro-Expansion Project, the largest construction project ever undertaken on the island. Though it promised to make the country virtually self-sufficient electrically, there are many critics of the project, including Clem. It has destroyed some of the pristine environment that makes the island unique. There was no environmental impact study, and the people of the island weren't consulted or asked.

Just downstream is another vocal critic, Anne Jno. Baptiste (née Gray) who, with her Dominican husband Cuthbert, owns the 10-room Papillote Wilderness Retreat at the base of Trafalgar Falls, the greatest tourist attraction on the island. Twenty-eight years ago, Anne fled Brooklyn to the island. After several enterprises, she settled into operating an intimate inn on the slopes of the volcano Morne Macaque, just below the Falls' shimmering twin plumes. There she has fashioned a private paradise, spectacular gardens filled with ficus, lilies, hibiscus, anthurium, cycads, heliconia, bromeliads, daturas, aroids, ginger, begonias, ferns, orchids, and poinsettia. Anne, though, is incensed, as the Hydro-Expansion Project may divert as much as 40% of the water from the main falls, lessening the beauty and perhaps the number of tourists who will visit. And, for the immediate future, her front lawn has been turned into a construction site filled with pipe and machinery. Almost everyone I met during my week on the island was enraged over the Hydro-Expansion Project, and it seemed the issue had become *the* subject of discussion.

Some of the project could be seen as we stepped to the mouth of Titou ("little throat hole" in Creole) Gorge. It was recently one of the island's hidden treasures, a narrow, sinuous slot canyon through which the intrepid could swim to a feathery waterfall. Legends of haunted chambers kept superstitious locals away, while its tough access kept the tourists at bay. Now, the basalt gorge was partially filled with rock debris from the digging upstream, and we passed a mother and her two young children wading to the falls for a picnic, all wearing "Don't Worry, Be Happy" T-shirts.

At the mouth of Titou Gorge a sign indicated the trailhead, and there we turned into the luxuriant handiwork of nature and began the trek up the flank of Morne Nicholls,

Opposite, top. A group of young Quadrille dancers in full Creole dress at the eighteenth-century Fort Shirley in the Cabrits National Park.
Bottom left. The official language of Dominica is English, though the lingua franca is "smiles."
Bottom right. A house in the Carib reserve, where the last descendants of the ancient West Indies race survive.

Opposite, top. A woman dressed in the "Wob Douilette," the Dominican national costume. *Bottom.* The great majority of Dominicans are Roman Catholic, a vestige of the period of French rule.

named for a nineteenth-century botanist who got lost (and was later found) on this mountain. It was dark and cool under the jungle canopy as we climbed the irregular steps carved from the fat trunks of elephant ferns. This was the dry season, yet we were always enshrouded in a fine-spun, swirling mist, what the Dominicans call "liquid sunshine." Mud was the most common ground. Not more than thirty minutes into the hike the skies opened, and we were pummeled with a hard, cold rain: a dose of the three hundred inches that the eastern trade winds drop annually in the mountains here, a volume so high that Dominica actually exports water to the drier Caribbean islands, such as St. Maarten.

But as soon as we pulled off our packs, pulled out the ponchos, and struggled to put them on, the tropical tempest lifted like a stage curtain, and shafts of soft light poured through the lacy canopy. The damp rain forest filled with scents of honeysuckle, lemon grass, and fermenting leaves. The variety of vegetation was astounding, sometimes sixty different trees and plants in as many steps. Long lianas hung from gnarled branches like the strangled snakes of Hercules. At one point Clem abruptly halted and we comically bumped into his back. He pointed into the riot of green that pushed against the trail, and not far away a white puff fluttered through dappled light: a wild dove. Above us we heard the lonely, three-note adagio of the rufous-throated solitaire (*Siffleur montagne,* or mountain whistler), a bird with a call like a sad clarinet, but we never saw the elusive creature. We did, though, see a tiny blue-headed hummingbird, and that gave us sightings of two of the 162 bird species on the island. That not counting the iridescent blue butterfly, bigger than a parakeet, that fluttered around Pam's camera as though daring her to shoot. We plodded onwards, in what was starting to seem like a Sisyphean quest. Every precipitous crest led to another, every ridge descended only to rise. Every moldering step we took we knew we'd have to give back.

Carib war dance.

After plowing through the fertile lower montane forest, we emerged into the transition zone, where we passed stands of *Wezinye montany,* the only native conifer on the island. Then it was into a cloud-wrapped elfin woodland, marked with trees stunted from savage tradewinds and dripping with mosses and liverworts. The path continued to twine, like the back of a viper. (This on an island that officially has no poisonous snakes.) As we stepped along a narrow ridge I found myself playing hopscotch to avoid hitting the puddles dotting the path. Clem turned and saw my game, and immediately scolded me: "Don't avoid the puddles!" He then pointed to a stained and faded piece of brown cloth tied to a twig at the edge of the path. Looking closer I could see that it had once been a delicate white handkerchief. Clem parted the branches. Beyond there was nothing but an abyss, a drop of unknown depth into some dense, green hell. Clem casually told us that the previous June a thirty-eight-year-old German woman had taken this trek, and hadn't followed the advice. She avoided the puddles, and while stepping around one she plunged over the edge to her death. Only her handkerchief remained.

No longer concerned about my new Nikes, I followed Clem as he splashed up and then down the ravine to the Breakfast River, one of 365 rivers (one for every day of the year) that drain the island. We stopped for a moment to drink the clear, cold water, then trudged upwards. "There is only one way to understand Dominica," writer Alec Waugh observed in 1948 in a caveat to future adventurers. "You have to see it on foot and by the hour." I figured by now we must be approaching complete comprehension, as we'd been walking for four hours. As we climbed higher the primeval dark green gave way to tawny browns and purples and the stunted scrub to spidery bromeliads.

At the highest point of the trek we paused on a blustery land bridge, one affording vertiginous views in all directions. To the immediate south above a Rousseaulike landscape loomed the peak of Watt Mountain, 4,017 feet high; to the north, beyond deeply scored valleys, mist-enshrouded Morne Macaque, 4,006 feet. But the real views were to

the east and west, where from this single roost I could see both the Caribbean and Atlantic oceans, the first time I had ever seen two seas at one stop.

To the west, plumes of charcoal cook smoke rose from the fretwork houses of Roseau. Beyond the town, resplendent on a sullen sea, was a fully bloomed Yankee clipper ship, one of the Barefoot Windjammer Cruises fleet. Because the Commonwealth of Dominica has no deep-water port, cruise ships have never made the call. The exception has been the shallow-draft Barefoot fleet, which regularly anchors offshore and motors its swinging singles to shore. On the black sand beach of the Coconut Beach Hotel, near Portsmouth on the northwest coast, all-night parties were the rule. Thus, tourist boat visitation has been relatively small, and the lack of tourists has kept the island comparatively chaste. There are no tacky tourist stalls, no hotels of any size, no restaurants of any star quality, no sugar-white beaches, no McDonald's, no Gucci boutiques, no casinos, bistros, offshore banks, golf or tennis clubs, and just one stoplight. But all this could soon change if the government has its way. A $7.1-million-dollar cruise-ship berth designed to attract the masses is under construction in Prince Rupert Bay; an international jet strip is proposed, and seeking funding; and several hotel and resort complexes are in the planning stages. All these would change the fortress built by Nature for herself, the place now called Dominica, forever.

Houses, called *caneys*, used by the Caribs. One is circular, one rectangular. (These drawings by Oviedo y Valdez, first published in 1535)

South of Roseau I could see the ghostly blue-green mountains above Soufriere Bay. I had gone diving there the day before with Derek Perryman, a former pilot who had quit the skies to open up his own dive shop. Derek had taken me to an underwater world of tropical fish and coral as spectacular as any I had ever seen, all in water as clear as windowpane. But the *crème de la crème* was a dive through the underwater hot springs, full of thermal activity related to the volcano I was now straddling. It was like swimming through a large champagne glass, with silvery beads of warm water streaming upwards like giant fizz.

Turning 180 degrees I could gaze at the more aggressive Atlantic coast, and even identify a profile of Martinique in the distance. Out of sight, to the north, was the Snake Staircase I had climbed the day before, and beyond was the 3,700-acre, eighty-eight-year-old Carib Reserve, where the last descendants of the ancient West Indies race were making their final stand. The Caribs were a tribe of Amerindians who canoed from the Orinoco and Amazon valleys in pre-Columbian times, and before that had migrated across a frozen Bering Strait from Mongolia. Settling here they devoured the indigenous Arawaks in an orgy of cannibalism. In fact, the word *cannibal* stems from the Arawaks' name for these bloodthirsty invaders. Indeed, if one believes Dr. Rochefort's journal of 1658, the Caribs extended their high-protein diet to à la carte choice: French being tastiest, then the English; Dutch were dull and boring to eat; Spanish unspeakably gristly.

Passing through the Carib territory I stopped and talked with Ann Timothy, the Parliamentary representative for the Reserve. She softly told me that the Indians resented Columbus, who claimed to have discovered Dominica as he sailed by on a November Sunday in 1493, though he never touched its shores or met the Caribs who had occupied the island for centuries. She also said some Caribs are indignant about perceived present-day governmental policies. They believe more state money is being spent to save endangered wildlife, such as the sisserou parrot, than the endangered Carib peoples.

Beyond this summit ridge we picked our way down a steep creek bed to the Valley of Desolation, a moonscape of steaming fumaroles and vents, percolating springs, and mineral creeks of inky black, chalky white, and turquoise water, and of guttural noises, as though the Earth were trying to clear its throat. It was an otherwordly, unsettled landscape, thick with the stench of sulphur and hissing gases. Gingerly, with light Indian-moccasin steps, we negotiated our way over a mat of mosses and lichens, across gray and

Opposite, top. Ringed with isolated beaches, Dominica is an island of peaceful seclusion in a sea of over-touristed motes. *Bottom.* A careful crossing at the base of the 200'-high Trafalgar Falls. The river is one of 365 on the island, one for each day of the year.

mustard pumice crust, between patches of miasmic vapor and tarns of superheated water. It was like walking on thin ice, only here if we broke through we wouldn't die from cold-water immersion, but from a hot lava bath. Beyond the Valley of Desolation came the Grande Soufriere Valley, steamier and uglier, scorched and prehistoric—a testament to the violence of nature. It was like hiking through a foundry.

Finally, just before noon, after ascending through seven climate zones, we rounded a yellow-tinged hill and looked down on the object of our desires: the Boiling Lake, belching sulfurous fumes in a pit seventy feet below. Milky grey and filled with thick steam, it looked like a witch's cauldron of lethal chemicals. My sunglasses and camera lenses immediately fogged, and while trying to peer over the inferno's edge I was blasted with an instant sauna. Then a breeze washed across my shoulder, and for an instant I could see the entire body of water, two hundred feet across. It looked different now, more alive, sinister, like the ravenous mouth of a fire-belching dragon. Toward the center, where the ebullition was fiercest, it seemed to be breathing. Then the steam thickened and covered the lake like a lid.

Intrigued with this natural phenomenon, I looked around for a way to touch its waters, and saw a creek that carved a passage on the north side. I scrambled down the bed to the lapilli-and-clay shore of the Macbethian brew. The place reeked of rotten eggs. Several years ago I had made the trek to the source of the Ganges, following in the footsteps of countless Hindu spiritual aspirants. They had made the pilgrimage to that holy fountainhead to take a bath in the lake that formed below the two-mile-high Gangotri glacier. I figured that was the thing to do here, so I stripped and eased my body into the water at the mouth of the creek. The water was shallow and cool, and I prepared for goosebumps as I slid deeper. Then it hit . . . scalding water. There was no transition zone, no buffer between temperatures. One inch it was 45 degrees; the next 190. Still, I was determined, and poked parts of my body into the lake water until I felt I had been thoroughly immersed.

Once out of the water, I dried off with my T-shirt and looked around the tiny canyon I occupied. It was a spectacular sight, like something from Earth's first morning, certainly unchanged since Dr. Henry Alfred Nicholls, the lost botanist, stumbled onto the lake in March of 1875. Then something caught my eye. I was directly below the normal viewing plateau for the lake, and in a corner I saw a candy wrapper, then a beer can, then a plastic cup. As I scanned the area I saw more and more trash. This site, so rare and magnificent, had been treated as a private trash bin for previous visitors. Horrific. It reminded me that just a couple days previous we had visited one of the other great natural wonders of the island, the Emerald Pool, the setting of a spectacular waterfall falling through the lush forest. From a distance it was a scene beyond belief, but as I walked to its base for a closer inspection I saw a tampon tossed behind its curtain, and a wall scarred with graffiti. Even the trees were marred, strips of bark pulled away from the bois bande trees by Guadeloupeans who believe that the bark is an aphrodisiac when chewed.

A hot, foul wind swept around me as I scanned the refuse in this refuge. A similar wind took hold of my emotions. What was wrong with this island that had so much, that had escaped so many of the environmental problems plaguing the world? It had thus far avoided the mistakes of neighbors, such as Haiti, who have suffered the effects of severe and wasteful development, soil erosion, vanishing of natural vegetation, and a marked change in climatic patterns. Was the future of Dominica the gross despoiling of its natural world, like that of Hawaii's Waikiki and Spain's Costa del Sol? With 80% of its ridged and rumpled land enwrapped in virgin forest, the island the Caribs called *Waitukubuli* ("Tall is her body") is one of the most pristine on the planet, and yet it seems headed down a collision course of pell-mell development that could destroy all that has

Opposite, top. There are few television sets in Dominica. This is one good reason why. *Bottom.* The endangered Sisserou parrot is the oldest species of Amazon parrot in the world and is found only in Dominica.

Overleaf. A waterfall in Morne Trois Pitons National Park.

Below. The Boiling Lake is the second-largest thermally bubbling body of water in the world (the largest is in New Zealand).

made it unique. "A wild place, Dominica, savage and lost . . . [it] will protect itself from vulgar loves," the late Dominican-born author Jean Rhys (1890–1979) had written. I wondered if her words would hold true to their meaning. As I put on my clothes I reached into my pocket and found a 25-cent piece. I tossed it into the Boiling Lake and made a wish, one I can't reveal or else it cannot come true—but it has something to do with my thought at the moment, and the integrity of a special place.

Before leaving I took one last look around the lacustrine basin. I had followed a trail of my own making from the sea at L'Escalier Tete-Chien to the greatest natural energy site on the island. I had stepped in the puddles along the way. I'd bathed in the seething waters. Yet I felt nothing, except a clammy wind from the brooding lake. If anything, I felt disappointed—not in the scenery, which was truly spectacular, but in the fact that such a place could be so carelessly violated with trash. As I put on my pack to leave, Clem offered me a soursop fruit to suck on; it was delicious, and mellowed my mood. I asked him if I should toss the rind over the edge, since as a tropical fruit it was eminently biodegradable. "No," he said, "I'll carry it out. We shouldn't leave anything behind." I nodded, and shouldered my pack for the long hike back. Maybe, I thought, this was the wrong site for the terminus of the Snake Staircase. Maybe I was too eager to fashion L'Escalier Tete-Chien into some kind of Rosetta stone for deciphering Dominica's consciousness—and my own. Maybe the full moon the day before disqualified me. Maybe the whole notion was just ancient-fable hooey.

Ten hours after we began we reached the van, and minutes later fought the five o'clock traffic as the workers on the Hydro-Expansion Project barreled down the corkscrew road in rattling open-bed trucks. When we reached the Fort Young Motel I was exhausted and muddy, and as I picked up my room key I was looking forward to a rum

punch and a lingering hot bath to ease my blistered feet and aching calves. But a message awaited at the desk asking me to immediately call Kenny Alleyne, general manager at the National Development Corporation Division of Tourism. It read "Urgent."

"The Prime Minister has agreed to see you in twenty minutes," said Kenny. "I'll pick you up in fifteen."

"Should I wear a coat and tie?"

"Nah. She's strictly informal."

Not even aching legs could keep me from this appointment. I had asked days ago if I might talk with Eugenia Charles, the world's only black female prime minister, and the lady orchestrating all the changes on this precious island. I had to make this date.

Minutes later, still dripping from a quick shower, I climbed into Kenny's car, and we made our way south to the Prime Minister's house. It was not the official state house; that had been destroyed, along with two-thirds of the island's structures, in Hurricane David in the late summer of 1979. The hurricane hit less than a year after the island received full sovereignty from England, ending more than two hundred years of colonization. We pulled up to her home, and it was not what I envisioned a prime minister's residence to be. It was a ramshackle, faded white summer house, not much bigger than a cottage. A single security guard with bad posture stood at the garage entrance.

Ushered inside, we found the seventy-one-year-old Miss Charles (she had never married) hunched over a messy desk in animated conversation on the phone. On the wall was the Dominica crest emblazoned with the national motto, *Apres Bondie c'est La Ter* ("After God it is the Earth"). When she hung up, she waved us to some chairs as she hobbled to a settee and flopped down. She had recently hurt her foot on a trip to Taiwan, so she wanted to talk while reclining. This was not how I pictured my first interview with a prime minister.

I asked innocuous questions for the first few minutes, uncertain of protocol, and she gave me the usual canned responses. Then Clem's words echoed through my head—"Don't avoid the puddles." I launched into the issue that had been bothering me since my arrival in Dominica.

"Ms. Charles, with its environmental and cultural integrity, Dominica is unique in the Caribbean. Why are you threatening this specialness with the Hydro-Expansion Project, the cruise-ship berth, the planned hotels, and the international airport?" I was calm and measured in my delivery, but eco-sense was raging inside, and I felt I had the lady cornered.

She paused. "You know, a few weeks ago I received a letter from a woman in Seattle who asked your questions, only she wasn't as polite. She thought I was destroying this island. What right did this lady have to say these things when she had only spent a week here?"

I nodded, and shifted a bit in my seat. I was just on my sixth day in-country. She continued.

"She doesn't know this country, she doesn't understand these issues. The Hydro-Expansion Project is the most ecologically correct thing we can do at this juncture. Right now 80% of our electricity is generated with diesel fuel. That's costly imported fuel that pollutes the air. The Hydro-Expansion Project can give us 100%, non-polluting, inexpensive power. And, I've been assured that it will not decrease the flow over Trafalgar Falls. Certainly the landscape is marred for the time being, but for the most part it is temporary, and it's a trade-off we have to make."

"What about the moves you're making for mass tourism development?"

"Again, you don't understand. The people of Dominica need to survive. They

"This Island [Dominica] is inhabited by the Caribbians, who are very numerous in it. They have a long time enertain'd those who came to visit them with a story of a vast and monstrous Serpent, which had its aboad in that bottom [Borlin's Lake]. They firmed that there was on the head of it a very sparkling stone, like a Carbuncle, of inestimable price; That it commonly veil's rich jewel with a thin moving skin, like that of a man's eye-lid; but that when it went to drink, or psported himself in the midst of that deep bottom, he fully discover'd it, and that the rocks and all about receiv'd a wonderfgul lustre from the fire issuing out of that precious Crown."

—John Davies, in *The History of the Caribby-Islands*, published in London, 1666

Various pictographs done by Caribs.

have every right to live as you or I. They deserve running water in their taps, electricity in their kitchens. Right now almost all the income for this country comes from bananas. We have a protected subsidy agreement with the United Kingdom. Even with this special arrangement we are one of the poorest countries in the world. Well, with the EEC creation of a single European market in 1992, the subsidy may disappear, and that could wreak economic havoc on Dominica. I'm looking for ways to avert that crisis, intelligent, prudent ways. I've had a number of talks about merging with the three other English-speaking Windward Islands, and giving up my position for the good of the whole. But I don't think that's going to happen for a while. I've been pushing for the farmers of Dominica to begin growing other crops, but we'll always have a hard time competing with the larger, flatter, more agriculturally efficient countries. And, I'm exploring tourism, which is a low-impact, non-polluting source of foreign exchange that provides jobs. I don't want 1,000-room hotels, I don't want package tours, I don't even want our own airline. I don't want tourist pollution, and I don't want visitors who never see what color our faces are. But at the same time I don't want Dominica to become a museum piece. I want to allow a little more tourism, but of a high quality. One personal goal is to make Dominica an international conference center for conservation groups. But for now we're looking to attract more nature and adventure tourism, and we're proceeding slowly, cautiously, with a goal of sustainable development. We had about 32,000 visitors last year. We're looking to increase that by about 15%. If the new airport is built, I've stipulated it be too short for wide-bodies. That would be too much, too fast.

"I don't want too much pressure to be put on our extraordinary environment; I want to preserve it. But I want everyone on this island to be able to feed and clothe his family, to receive an education and health care. What I want is the correct balance. And that righteous Seattle lady, who went home to her VCR and halogen lights, leaving us in our candlelight, just doesn't understand."

A piercing look punctuated this last point, and the setting sun threw arrows of light through the jalousies of the book-lined room, painting her cheek red. "One must wait until the evening to see how splendid the day has been," Sophocles had said. As I watched the light illuminate the woman on the couch I thought that indeed the day had been splendid.

We continued our discussion past the appointed hour, until we were interrupted by two directors of the Strom Thurmond Institute at Clemson University, there to talk of investment. In her ten years as prime minister Eugenia Charles had weathered coup attempts (including one organized by Ku Klux Klan members and neo-Nazis from America), devastating hurricanes, crop failures, international condemnation for her role in the U.S. invasion of Grenada, and severe economic strife. Yet perhaps none of these was as challenging, as consequential, as the current battle to achieve the right balance between the environment and the needs of her people. I liked the Iron Lady of the Caribbean. I wasn't sure I agreed with everything she said, and knew that as a lawyer and a political survivor she played the public relations game with skill. But I felt her stated convictions were sincere. I felt the same of Clem and others I'd met on the island, and I felt the operative word was balance. There are no extreme or simple solutions in safeguarding the environment, in delivering humankind. There is only a delicate, perilous balance among all life. The landscape is no longer innocent. Though we inherited some flowers from the Garden of Eden, the trail of the serpent is over them.

The puddle had started to clear. As exhausted as I was, I felt the long and winding trek had been worthwhile. That night, as I packed to leave, I looked out the window. There was the moon, which still looked full. When the clouds passed in front, I couldn't help but think it looked like a distant boiling lake, or the gaping mouth of a giant snake.

Newfoundland

The Mirage of Wilderness

We were flying over a desolate expanse of rocks and great pools. "Bloody country," he ejaculated.
—Lord Moran, quoting Winston Churchill
December 2, 1953

It is impossible not to confront the Janus nature of this land.

Its rivers run clean and clear; its air is alive with the breath of honeysuckle. Its forbidding interior is a land without litter. Its craggy edges and joints are lodged with deeply religious Protestants and Roman Catholics renowned for their charity and moral excellence. On a sunny day the place seems like heaven. Yet the incessant storms, the frigid waters, have snuffed countless lives, and a pall of violence forever hangs. The people here are children of their beloved enemy, the sea. They move with the rhythm of the natural world. Often characterized as optimistic fatalists, they are a thick-skinned and gentle stock who exterminated the Beothuk Indians, hunted the great auk to extinction, and brought the pilot whale to the brink. For 450 years the economic mainstay was a seemingly inexhaustible supply of saltfish. Now the stock has been reduced to a tiny fraction of the glory days. Not long ago chief livelihoods included clubbing young seals to death. Presently they include mining, damming wild rivers, and felling trees. Despite these intrusions into its wilderness, few places survive with such environmental integrity and harmony; yet the urban-based environmentalists of the planet have painted The Rock as a house for eco-bandits.

The island is Newfoundland.

I wasn't looking where I was going. Instead my eyes were sweeping the sky, caught by the sight of endless skeins and clouds of flickering wings. It was a world of birds, millions of them: arrow-swift murres, Pillsbury dough-bird puffins with their clown-colored beaks, great-winged gannets flying arabesques betwixt and between until the sky seemed alive with flight. I continued to paddle as I ogled the phalanxes, until suddenly I heard a thud. It was not more than a tap, really, much like the force I imagined a Newfie sealer used when clubbing the nose of a whitecoat. Looking down I saw I had bumped broadside into Gerald's kayak. My heart sank as I saw him teeter back and forth, and then in a slow-motion roll, he poured into the deep indigo-blue waters of the icy North Atlantic.

I remembered the words of our guide, Jim: "A person can only function for five minutes in this water; then he goes numb, helpless." I positioned my kayak next to Gerald, and reached to him as he grabbed the edge of my boat, almost turning it over in

Opposite. Bonne Bay in Gros Morne National Park, Newfoundland.

Opposite, top. The Dragon's Throat, a grotto the bow of a kayak can tickle, on Great Island in the Witless Bay Ecological Reserve.
Bottom left. A world of wings. Northern gannets, Newfoundland's largest seabirds, in Cape St. Mary's Bird Sanctuary.
Bottom right. Salmon running up Big Falls on the Humber River. Some of the best sport angling in eastern North America is here.

panic. This wasn't right, I thought, and rocked my hips to maintain balance. We were both saved by a sharp command from Jim: "Richard, move out of the way." With my paddle I dug a few strokes towards the spume of a humpback whale several hundred yards away, then turned around to watch. With cool, quick professionalism Jim and his protégé, Young Doug, lined up on either side of the overturned kayak. Placing a paddle between the upright kayaks as a brace, they pulled Gerald's boat from the water, drained it, flipped it over, and positioned it in the water between them. Then Jim instructed Gerald to pull himself on board as the guides stabilized his craft. In four and a half minutes Gerald was back on board, the rescue a vivid memory, and we continued our paddle to Great Island.

Those who live here call it The Rock or the Granite Planet. The explorer Jacques Cartier christened it "the land God gave to Cain." None of these does it justice. A garden of wildflowers, a sanctuary for moose and caribou, host to the greatest fish pastures in the world, a landscape of tall trees and hard splendor, Newfoundland is much more than a slate stopper thrust into the bell-mouth of the Gulf of St. Lawrence. It is the tenth largest island in the world. Over 1,000 miles northeast of New York, it is the most easterly land in North America. It turns its back upon the Canadian mainland, barricading itself behind the 300-mile-long rampart that forms a western coast as tattered as an old fishing net. Its other coasts all face the grim ocean, and are so slashed and convoluted that they present more than 5,000 miles to the sweep of the "Big Pond," a favored Newfie name for the Atlantic. Newfoundland is of the sea, and so I had felt there could be no better way to explore the narrow gaps bitten into its foreshore—its coves, bights, inlets, reaches, runs, and fjords—than by sea kayak. So it was I found myself with Canadian Canoe Adventures travelling through the glacially scoured scapes that define an old land, one that some insist has yet to be found.

Our week-long sojourn started on Friday the 13th at Lower Lance Cove on Random Island off the serrated southeastern coast of Newfoundland. The first thing Jim Price, our guide, did was go through our gear and winnow out 75% to be left behind. "This is not a cruise or a raft trip," he said, his eyes crinkling at the corners as he spoke. Besides, as I later discovered, he wanted as much premium space as possible to pack his wife, Margie's, cooking, and for good reason.

When I was properly shaken down to a single change of clothes, I slipped on my sprayskirt (backwards at first), and slid into the sleek and form-fitting twenty-foot blue-and-white fiberglass Seascape tandem kayak. This craft has come a long way from the driftwood-and-animal-skin version devised by Eskimos four hundred years ago. I would be sharing the boat with Pamela Roberson, the photographer for our expedition. This was my first time in such a craft, and for the first few minutes I felt like a goose among swans. Finally, though, I got my sea legs and arms and we were off, the bow shedding waves as we paddled east down Smith Sound towards the North Atlantic.

The boat was remarkably stable yet agile, cutting through the water like a slim missile, and after an hour I felt as though I'd been born into it. The vexations of the urban, managed world washed away with the water that fell from my paddle blades, and I expanded my chest in the big cold freedom of salty air.

Among the icebergs, off the coast of Newfoundland. (from *Harper's*, March 1861)

The scenery was exquisite. We cruised along a ripsaw coastline marked with black spruce, stunted fir, and gaunt granite walls, past the occasional lobster pot. Arctic terns by the hundreds spread their sharp chevron wings above us. A tiger swallowtail, Newfoundland's largest butterfly, fluttered across the bow. After a few leisurely hours we turned around a barb in the land, Hayden Point (after the captain who wrecked his schooner there), and paddled into a small, protected bay called Gabriel Cove. We were at Thoroughfare, a once busy outport, now abandoned. All that remained was a pelt of tawny grass swept with wild irises, purple lupines, tall meadow-rue, marsh marigolds, daisies, and buttercups.

Here we set up shop, and hiked into the birch and tuckamore to the ruins of a once active merchant post that served the "thoroughfare" tickle through which boats travelled to and from Smith Sound. Today, picking through the planks and wood chips, we could identify just two former structures: the town sawmill and the steeple of the church. After our little archaeological dig we washed and collected soft water from the little spring that trickled down an alcove just across the inlet from our campfire, and after a dinner of fresh salmon and cod tongues fried in batter we made an early retreat to bed.

The next morning, after a nursery-colored sunrise and the alarm clock of a robin singing near my tent, we sat down to a breakfast of pancakes, sausage, and thick coffee. During our second helping we looked up to see what appeared to be a black-bearded pirate approaching in a white kayak. It was Mark Dykeman, Jim's partner, just in time for the second pot of coffee. Working as a construction manager in St. John's, he could only take the weekend off, so he had left at 4 A.M. this Saturday morning and kayaked the nine miles to meet us for breakfast.

By mid-morning, under a sky that foretold rain, we were off and paddling across the tickle towards another, smaller island, Ireland's Eye. To get there we had to leave our protected cove and round an exposed stretch of the island, a stretch lashed by the waters of Trinity Bay. For a moment I felt we had dialed back a thousand years and were part of that small band of Norse explorers steering high-prowed longboats up to the New Founde Land. But my Viking fantasy lasted only a brief moment. Just as we broached the rolling waves, my feet slipped off the pedals controlling the rudder. I squirmed around inside the boat, trying to reposition my legs and feet, but couldn't make the purchase. A wave of panic rolled over me, and Pamela began to yell at Jim, paddling merrily along a few hundred yards ahead. After several screams and some frantic paddle waving he saw us, and sprinted back to our bobbing boat, which had turned into the wind and was weathercocking. In a flash he ripped off my sprayskirt, reached into the bulkhead, readjusted my rudder pedals, and resealed my skirt. His face wrinkled up in a cocky grin as he motioned for us to follow, and off we went.

After a few miles we turned into the snug harbor that once served the town of Ireland's Eye. It was called that because a hole in the rock faced towards the Emerald Isle, and legend held that on a clear day a viewer could spy the ancestral home through the hole.

As we negotiated this harbor it appeared we were paddling back into the seventeenth century, which is when the town first appeared on maps. Big chimney-potted clapboard houses with mansard roofs and curved dormers lined the cliffs on both sides. Directly in front of us, at the end of the bay, stood a large, white, wooden, neo-Gothic Anglican church. But the windows had no eyes here; the pews gave forth no songs. Ireland's Eye was a ghost town now. The only living creature we saw was a great bald eagle who swooped over us, glowering with amber eyes, signalling, it seemed, that he was now the mayor and constituency of Ireland's Eye.

The day turned mauzy, a popular Newfie word meaning cloudy and wet. We parked in front of a blossoming lilac tree, and under an oblique rain Jim brewed a pot of tea and cooked up a pithy stew packed with pieces of a moose he had shot months before. After a dessert of Margie's cookies (best I've ever tasted; I wished I'd left even more gear behind and made more room for Margie's baked delights), we trudged up a dripping-wet overgrown path across a landscape that seemed to have risen overnight from the sea, and began to explore the burg of Ireland's Eye. The town was one of 148 communities in the Random Island area resettled in the 1960s by a purblind government in an effort to centralize the population in "growth centers," where public services such as transportation,

Below. A gothic wooden Anglican church presides over the haphazard scattering of small, tidy houses of a Newfie Outport village.

schools, and medical care could more readily be provided, and the island could be recast into an industrialized principality. Over 20,000 people were promised new jobs and a better life, and coerced to move as part of this misguided program that saw Newfoundland turning its back on the ocean and becoming a neo-Detroit. By and large the jobs were made of air and sea foam, and the new life was one of psychic and spiritual havoc.

Now the town is most famous for its drug bust. In August 1988, local fishermen, suspicious of high-speed boats zipping in and out of Ireland's Eye, called the Mounties, and Canada's largest hash bust took place. Sixteen tons, with a street value of $200,000,000 Canadian, had been stashed inside the cavernous church where we now stood. We admired the chancel and apse, and wondered how it could ever have been forsaken. Then we hiked back behind the church a few hundred yards through a thick alder grove, and stepped through a spindly white fence to the graveyard. Woodland mushrooms lined the raised edges of the graves. The tombstones, slanted with the settled soil, were decorous affairs, carved in white marble, and I could clearly read inscriptions that told of people born over two centuries ago: EDWARD COOPER, DIED MAY 6, 1868 AT THE AGE OF 87. I wondered of his life and of life in his time, and felt certain it was little different a hundred years later when his relatives were forced to relocate and abandon their heritage.

That evening, after a dinner of fresh mussels, Newfie steak (fried baloney), onions, and baked potatoes, Jim changed to his roll-necked guernsey and pulled from his haversack a bottle of Screech, an imported Jamaican rum so named for the reaction its potency tends to produce. After a few swigs, Jim and Mark loosened up and told us a little of why they entered the outfitting biz. Jim insisted it was "just for the halibut," while Mark confessed his goal was to someday turn outfitting into a full-time profession. For now, though, the motto of their tiny company was "Don't quit your day job." Both were keen kayakers, and had met seven years ago boating some of Newfoundland's wild rivers. They continued to spend weekends and vacations together exploring new waterways, and once even kayaked to France. Well, they kayaked the fourteen miles to St. Pierre, the French territory just off the bony Burin Peninsula, a southern finger of the hand-shaped island. Now, both in their early forties, they had decided to see if they could make a business as one of a handful of adventure tour operators on the island, and we were members of one of their first commercial tours.

On the way back from Ireland's Eye, Pamela's eye noticed a moving brown speck on shore. We paddled closer, and the fuzzy shape sharpened to a distinct form: a young moose. It was one of the 70,000 or so that now roam Newfoundland, a nonindigenous species introduced to the islands in 1878. It didn't acknowledge our presence, and continued to munch on partridgeberries, even as we parked the kayaks on shore just a few yards from his lunch spread. It was only when Pamela stepped ashore to get a closeup shot that the moose decided to move back into the boreal forest, back into the interior to places still unknown to reasoning animals.

The sun shimmered as though dipped in a bowl of crystal as we packed the following day, but it belied the task ahead. It was a grueling five-hour return paddle to Lower Lance Cove. When at last we pulled our boats on shore we met two plucky white-haired women, Blanche Ivany and her cousin Martha Stone. In a burr rich as Irish cream they told me why there were there. Blanche, a widower, was born in a four-square, two-story house in Ireland's Eye in 1931, and lived there with her fisherman husband, Lambert, until 1963. Then the government withdrew funds for the post office, the school, and the government store, and residents were inveigled to move. She and her husband were given nothing for their land or home, just $600 in expenses to reestablish within the Trinity Bay area. But once moved they could find no work, so Lambert would use his dory to make the long trip back to his old fishing grounds at Ireland's Eye. He died in 1987 at the age of sixty-

Opposite, top. The calm of this cove belies the violence always ready to blow in from the "The Big Pond," a favored term for the North Atlantic. *Bottom.* Our campsite above La Manche, an Avalon Peninsula outport abandoned in 1966 when a tidal wave washed away most of the town.

Canoeing among icebergs. (from an 1885 issue of *Harper's New Monthly Magazine*)

Below. 70,000 moose roam Newfoundland; they are a non-indigenous species introduced to the island in 1878.
Opposite, top. Kayaking around Great Island among a blizzard of birds.
Bottom. A herd of barren-ground caribou cooling off on a snow patch in midsummer in the high country of Gros Morne National Park.

one, dead of a broken heart. Now, every Sunday Blanche and her cousin, who was also a victim of relocation, came down to the shale beach at Lower Lance Cove and looked across the water towards their old home.

The next day we were paddling off the windswept eastern coast of the Avalon Peninsula, as far east as a paddler can get and still be in North American waters. We were in the Witless Bay Ecological Reserve, on our way to Great Island, one of the world's largest puffin rookeries, when I bumped into Gerald's kayak and sent him into the brine. Gerald took it all in stride, but I couldn't help but feel he hoped for some sort of revenge. It came minutes later, as we paused in a clangorous cove at Great Island to gawk at the loud, wheeling masses of beer-bellied puffins, black-legged kittiwakes, stubby-winged guillemots, cannon murres, and yellow-headed gannets, Newfoundland's largest sea birds. I had never seen such a sight. The sky was full of wings, and as I looked skyward I inadvertently opened my mouth in awe, and immediately felt something foreign drop in. It was bad enough to think of what had happened, but I hate anchovies, and could tell that's what my bombardier ate for breakfast. As I spat and wiped myself clean, Gerald's laughter rose above the din as though he knew the gods had just evened the score.

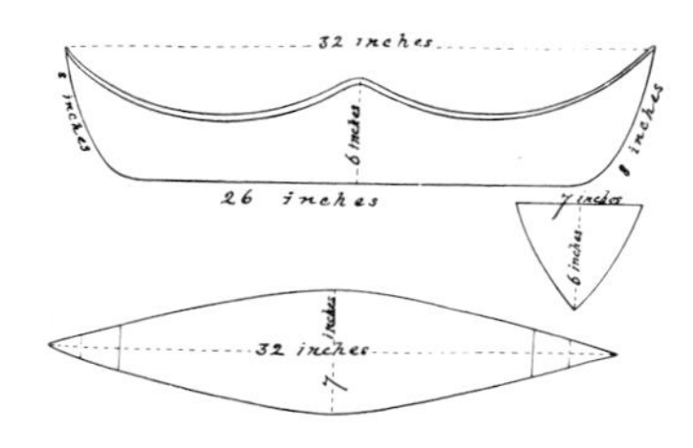

Beothuk Indian canoe, 1915. As islanders, the Beothuk depended upon marine resources and used birch bark canoes. They travelled forty miles to the Fink Islands to gather eggs of seabirds.

We circumnavigated the much-guanoed Great Island, poking into a sea-carved cave Jim named the Dragon's Throat for its esophageal rumblings, and running through a natural sea arch. Then we paddled back to the mainland, towards the town of La Manche. It was another abandoned outport, but this time not because of political gerrymandering, although there had been government pressure to move. In January 1966, a tidal wave had washed away all the boats and stores of La Manche, and most of the homes. It was as though God were siding with the government, and so the residents of La Manche capitulated and agreed to be resettled.

We actually beached at the wharf at nearby Bauling East, an active fishing community with a knot of brightly colored homes. Leaving the protected bird sanctuary behind, I paddled with a feeling that Newfoundland was perhaps now sailing with the right environmental tack. Then, as we pulled our boats onto the dock, I couldn't help but notice a stench mingled with the iodine tang of kelp. Dozens of dead puffins, their already dumpy figures looking even more bloated, were floating in the tidewash. They weren't the victims of insecticides, or oil spills, or poaching. Rather, of the cod fishermen's dragnets, which are spread over such a long distance that when hauled in they invariably capture a few of the puffins floating on the surface.

Left. Western Brook Pond is actually a rock-girt fjord. Its glacially scrubbed granite cliffs rear 2,200′ high.
Top. The lighthouse at Lobster Cove Head overlooking the Gulf of St. Lawrence.
Above. Sea gulls pluck these prickly, globular urchin fish, then drop them, like bombs, on our campsite at Gabriel Cove.

For the final leg of our kayak exploration we moved to the other side of Newfoundland, to the 450,000-acre Gros Morne (Big Gloomy) National Park on the primordial coast of the Gulf of St. Lawrence. Named a UNESCO World Heritage site in 1988, the park has been called "the Galapagos of plate tectonics" because of its exposed expanses of the earth's mantle and various stages of the earth's evolutionary history. The park also showcases Newfoundland's land conservation policies at their best: an extraordinary wilderness where moose, caribou, and black bear can wander completely protected, and people can explore a backcountry without crowds and trash bins. The park is even willing to "trial balloon" new policies that might help maintain the park at a savings to the government. Before getting back in our kayaks we took a hike up one of the park's more popular trails, Berry Hill, a rock knob that was an island during higher sea levels just after the last ice age. At the trailhead was a deep box filled with gravel, and a sign inviting us to participate in "an experiment in reducing costs while maintaining quality." Then it gave directions, asking hikers to fill the can with gravel, carry it to the top of the hill overlooking the park, spread the gravel on the trail as an erosion preventative, and then return the can at the end of the hike. It was an idea brilliant in its simplicity. The only problem was there was no can; it had been stolen.

Why Indians were called red: The Beothuk decorated their bodies with red ochre and were given the name Red Indians by white explorers. Reports of people with red skin coloring travelled through Europe, so the stereotype developed and grew that the New World natives had red pigment in their skin—redskins.

It was a two-mile peat-bog trek to the put-in at Western Brook Pond. Along the way we passed several insect-eating pitcher plants, the provincial flower. Once there, the clouds darkened, the wind began to wail, and the water whipped itself into whitecaps. This did not look like pleasant paddling weather, and even Young Doug, who had recently placed seventh in the flatwater kayaking competition in the Canadian Games, wondered aloud if the "white horses" (whitecaps) might be too rough for our planned trip. But Jim, fearless as a snake charmer, would hear nothing of it. Bubbling with boyish optimism (a result of Margie's cooking, I was certain), he had us launch and begin paddling against the cutting wind up the famous pond.

Western Brook Pond is not really a pond. In typical Newfie understatement, almost any lake or large body of water is called a pond. "To a Newfie a lake is a hole in your boot," Jim told me. But this was more of a rock-girt fjord, looking like something out of Norway or New Zealand. Ultimately it didn't matter what you called it, pond, fjord, lake—at that moment it was a combing sea of spindrift, and I was paddling in fear of a capsize. Yet, as I strained toward its gates, my mind fogged in fear, I couldn't help but look up and be stunned by the scenery. The cleft in the mountains ahead looked as though it had been split clean by a giant axe. It was like paddling into a flooded Yosemite Valley, one with no hotels, no galleries, no laundromats, no tourists. Just towering glacially scrubbed granite cliffs, 2,200 feet high, beckoning to me. I responded, not just for the view, but because between those protected cliffs I could see the water was a lot calmer.

An hour later I pulled off my poogies (elbow-long neoprene gloves). We were in the sheltering arms of the beetling cliffs, the wind now at our backs, and the water, while not Formica smooth, was at least forgiving. From here the ride was a pure delight. I even tried sailing, furling my lifejacket between the blades of two paddles, but the irregular wind made it a tricky and tiring endeavor, so I abandoned the technique for more traditional locomotion. We passed the dramatic spill of Blue Denim Falls on the left; gneiss hanging valleys on the right. Then Wood Pond Falls on the left, a cascade falling over 1,500 feet. The dark color of the water beneath us was an indicator of its depth, some 540 feet, and the water was Arctic cold, 49 degrees Fahrenheit. On the right, if we squinted, we could see the red granite seams in the face of an ancient mariner etched in the cliff top; a very old man indeed, probably around 600 million years old, if the geologists are correct. We stopped at a rookery of great black-backed and herring gulls, perched saucily on the cliffs. They took flight and scolded us for disturbing their day. I was careful to keep my mouth shut while admiring their aerobatics. Then, with a few more strokes, the canyon made a scimitar turn and we were faced with the *pièce de résistance* of the park, Pissing Mare Falls, the longest and most spectacular falls in Canada, dropping 1,850 feet from the canyon rim.

After ten miles of paddling we pulled into the pebble beach at the west end of Western Brook Pond. Not surprisingly, we were the only campers in this quasi-paradise. Quasi because though the beauty was nonpareil, the experience was a bit tainted by the blackflies—thousands of the pesky biters, always ready for a piece of exposed skin, or an available orifice. But we discovered the cure—Screech, in large doses, taken internally. After a few applications we didn't feel a thing . . . until the next morning.

It was late morning by the time we started up the trail. The plan was to hike to the rim of Western Brook Pond, where one of the grandest (and least seen) vistas in all Newfoundland could be savored, and then to return in time to kayak back before dark.

I decided to start out ahead of the others so I would have time to take photos. The trail was unlike any in a national park I had ever seen. A few yards from our campsite a rotting hand-carved sign pointed westward, the wrong way, and bore a simple designation: TRAIL. It would be the last sign, sure or otherwise, of our whereabouts. The path quickly vanished in the spongy tuckamore, but I didn't panic. There was only one way to go—up the U-shaped valley toward the Precambrian walls of the Long Range Mountains, the northernmost extent of the Appalachians. Behind was the pond, and on either side the sheer walls. Sharp and sudden as the side of a box, they defied negotiation. So, swatting through the brush, climbing up the boulders, wading through the muskeg, I continued upwards.

Below. Gros Morne means "Big Gloomy," and was named a World Heritage Site in 1988 for its unique exposed expanses of the earth's mantle and various stages of the earth's evolutionary history.

It was an intoxicating hike. Waterfalls materialized out of riven rock, and the views became grander with every step. This was July, yet patches of snow lit up the gray fans of scree. After a couple of hours scrambling up a recent rockslide, I was faced with a decision. To the left was a side canyon that looked as though it offered a passage to the top. The alternative was to drop back down a saddle into a second stream bed, and climb up the other side of the main canyon to a slope that appeared gentle enough for a summit attempt. I chose the closer route, the left ravine, and began my assault.

It was quickly apparent I was on the wrong route. The vegetation became denser; there was no way to move forward save to claw upwards, swimming upstream in a river of birch and springy juniper. It was grueling, sweaty work, but I kept going, feeling it the wisest course, believing I would be above the tree line soon, and then just a dash away from the top.

Shananditti, last of the Beothuk, circa 1825. She was born about 1800 and captured in 1823 by white bounty hunters. After serving five years as a domestic, she was discovered by William Comstock, a Newfoundland pioneer, who had been searching for the Beothuks. Before her death by tuberculosis in 1829, Shananditti related most of what is known of the Beothuks to Comstock.

After an hour of hand-to-hand combat with the goblin forest, of scratching through a tangle of larch and alder, I emerged above the tree line onto a glacial drumlin and surveyed the landscape. It didn't look good. The gully I had hoped to scale narrowed into a dark chimney, and the final pitch of 100 feet or so was slippery and sheer, impossible to traverse without ropes and pitons. I had successfully climbed to a dead end. Then I heard Jim's voice echoing from across the canyon: "Where are you?" I yelled back, and thought I could see a rustle of trees about a mile down the abrupt valley. "Come down!" Pamela's voice now reverberated. But I was exhausted, and needed a few moments before I could move, so I sat down on a lichen-covered rock and pulled a Mirage chocolate bar from my pocket. As I unwrapped it I looked back down the valley for the first time in a couple of hours, and was stunned by the sight. Some blessings come from nature unbidden and unplanned, and this view was one. I imagined it the sort of landscape the Maritime Archaic people who made their way here along the edge of retreating glaciers 10,000 years ago must have beheld. To my left was the misty veil of the great falls. Above me was a bald plateau where I could make out a small herd of barren-ground caribou cooling off on a snow patch. I could gaze all the way to the end of the snaking Pond, and beyond to the Gulf of St. Lawrence, the waters infamous as the killing fields for millions of baby seals. It was tranquil now, deep blue—no sign of the red tide, the blood of countless seals that had so recently stained these waters.

As I sat there watching a scene of unearthly calm, I munched my Mirage bar and reviewed the lineaments of this odd island. When I finished the candy I bunched up the wrapper and began to stuff it in my pocket. Then I stopped, balled the wrapper tighter, and tossed it against the cliff. Somehow it seemed the thing to do, like Ed Abbey's insistence on throwing beer cans along the highways that dissected his sacrosanct desert landscapes. But this wasn't out of anger. It was to defy the immutable morality of the environmentalists who had never visited this place, but had so heartily condemned it. I knew the simple act of leaving trash in a wilderness would incur the wrath of anyone with green leanings, and I counted myself in that troop. But in this case it didn't warrant it, or so I rationalized. I was probably the first human to ever stand on this perch. Even the Beothuk Indians wouldn't have been stupid enough to try this route; likely no other human would for many years to come. By that time the rain and storms and severe climate would have long obliterated the paper wrapper. If a tree falls in the woods and nobody is there to hear it, is there sound? If a wrapper is left where nobody ever sees it, is that trash? Perhaps.

The small act also seemed in some way to express my frustrations with the ecological invective hurled at this island and its people. There seemed a singular cohesiveness of culture and society here, and a unity with the natural world. I deeply admired the Newfoundlanders' famous traits, traits worn on the sleeve of Jim Price: self-sufficiency, adaptability, daring, absolute endurance, unbounded hospitality, a rare concern for fellow man, an appreciation of wilderness, and an evergreen goodwill that triumphs over the futility in life. The history of the people here is one of foreign exploitation and interference in the modest goals of feeding and sustaining a healthy family. For centuries they had caught or killed what the Great Waters had offered them: seals, whales, and codfish. But then the world turned against them, condemning the hunting of seals and whales. At the same time foreigners were employing high-tech vessels with sophisticated radar to beat them in the fishing game. In 1986, when Spain and Portugal joined the European Community, they blithely ignored the voluntary fishing allocations, and every year since have taken five times their quota from the shallow waters just beyond the 200-mile Canadian limits. The fall in fish stocks forced the Canadian government to cut its own 1990 quotas by enough to throw 3,000 Newfoundlanders out of work, and scientists insist quotas will have to be cut far more deeply if stocks are to revive. This in a province where the official unemployment rate is 17% (privately some told me it was in fact closer to 30%), the sales tax is an ungodly 12%, and incomes are only two-thirds the Canadian average. In recent years many former sealers, whalers, and fishermen have turned to logging, but now a vocal band of outsiders is again jeering the destruction of a limited resource. Now everything seems to be running short except hard-luck stories.

Below. The Gannet Stack on Cape St. Mary's, where the distinctive seabirds with their yellow heads and black wingtips gather and flock. *Bottom.* The legendary Newfie character includes an evergreen goodwill and an unbounded love for family and pets.

Nonetheless, the resourceful are turning to new sources of income. I couldn't help but notice on my journey that the island highways are lined with cheap motels, gas bars, and waterslide parks (when there are less than sixty hot swimming days a year). Tacky tourist shops sell mock cans of moose, ceramic Newfoundland dog decanters ("produced entirely by local craftspeople"), lobster parts glued together to look like a fisherman, "In Cod We Trust" posters, and Newfoundland In A Can. In 1988 the government invested over $17 million Canadian in a cucumber greenhouse scheme that quickly went bust. And while I was there the Economic Recovery Commission announced it was going to invest in an ice factory for the exportation of Newfoundland ice. And then Jim and Mark have their kayaking adventures.

There is no pat answer. As components of a vulnerable living fabric we cannot allow the destruction of any species, but it is too bad the human side of the issues are rarely adequately addressed. Greenpeace, Brigitte Bardot (who made a well-publicized trip to an ice floe in 1977 to protest against the annual seal-pup hunt), and animal-rights author Farley Mowat ("Hardly Knowit" is a favorite local nickname) are practically national enemies in Newfoundland; they portrayed the good people of The Rock as biocidal Darth Vaders, when in fact they are decent and in many ways extraordinary folk simply trying to eke out an existence in ways honorable just a few years ago. The key is to find viable alternatives, and the environmental groups would do better to back off the personal censure and castigation, and work with the Newfoundlanders to build a better life, something beyond the Rod and Gun Waterslide Park.

Spearing a shark. (from *Harper's*, 1861)

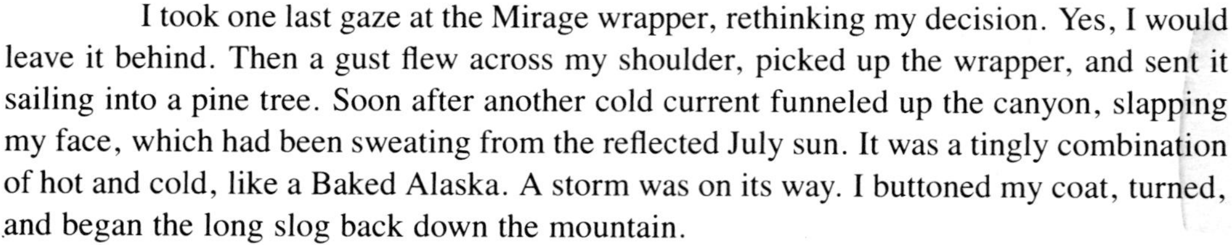

I took one last gaze at the Mirage wrapper, rethinking my decision. Yes, I would leave it behind. Then a gust flew across my shoulder, picked up the wrapper, and sent it sailing into a pine tree. Soon after another cold current funneled up the canyon, slapping my face, which had been sweating from the reflected July sun. It was a tingly combination of hot and cold, like a Baked Alaska. A storm was on its way. I buttoned my coat, turned, and began the long slog back down the mountain.

Three hours later we were recongregated at camp, battening down the hatches. A squall had grabbed my Eureka tent and tossed it into the hemlocks, and other bits of raiment and light gear were scattered with the wind. The waterfall above us was actually being blown upwards, so the water seemed to be running upside down. On the plus side, even the blackflies had been blown away. The pond that had looked so composed just a few hours earlier was now a stark and wild inebriate. "We've got gale-force winds; we can't kayak. We're stuck," Jim announced in a sober tone that was completely unfamiliar to this point. It looked as though we would have to hunker down for the night and wait out the storm, when around the bend appeared the *Westbrook 1,* the 42′-long skiff that regularly carries tourists up to the end of Western Brook Pond. Pamela took out a white windbreaker, attached it to the top of her paddle, and waved at the tour boat.

Skipper Charles Reid steered his pitching vessel towards our encampment, then announced over his loudspeaker that it was too rough. He couldn't make it in, so we were on our own. It was daunting news, but then the russet-faced skipper turned the boat around and somehow backed into our little cove. Quickly we threw on our kayaks and camping gear, and were soon sailing back to safety at the eastern end of the pond.

As the *Westbrook 1* cruised amidst the skirling wind and water spouts I climbed topside to take a last look at this uncommon landscape. The canyon was filled with blue tendrils of fog, and cold water sprayed across my glasses, but for a moment the mist cleared and the sun shone through, lighting up the brooding cliffs and the grand falls, turning its spray into a brilliantly wavering spectrum of color. It was a magic moment, and I couldn't but think that the entire scene looked like a mirage.

Bahrain

Under the Volcano

The croak of the raven is not heard,
the lion does not devour, the wolf does not rend the lamb,
the dove does not mourn, there is no widow, no sickness,
no old age, no lamentation.

—ANONYMOUS DESCRIPTION OF THE ISLAND "30 DOUBLE HOURS SOUTH OF BABYLONIA, IN THE MIDST OF THE SEA OF THE RISING SUN," THIRD MILLENNIUM B.C.

It was a legendary place called Dilmun.

The Sumerians, who hailed from what is today southern Iraq, described the island of Dilmun as the Garden of Eden, the lush paradise where Man's innocence was forever lost. This dry patch of desert was where, on a hot June morning in 1932, a steel bit pierced a layer of blue shale two thousand feet down, and a jet-black torrent erupted. It was the first oil discovered in the Middle East, the beginning of an age. And there was a sense that at this moment, this was where it could all end.

The whole thing seemed the worst of 1,001 Arabian nightmares. Why hadn't I listened to reason, or at least to neighbors and friends? It was 10:00 at night. I was standing in the immigration hall at the Bahrain International Airport after twenty hours of flying, and the official told me there was no visa waiting, no one to meet me, that nobody had ever heard of me. Sorry.

Great. I was "hidden like a fish in the sea," as the Sumerians described the island of Bahrain around 2500 B.C. I had no names, no phone numbers, just stateside assurances that everything was arranged, that I would be a guest of the royal family. I felt I was lost in an algebraic equation that might or might not be solved. After an hour of wandering about, asking advice of policemen, passersby, the cleaning staff, I decided to call the man who had set up this visit, Mohamed Hakki in Washington, D.C., the former press secretary to Anwar Sadat. I got through in a snap, and the connection was so clear I though Mohamed was in the next room. He gave me the name of the Prime Minister's press secretary, Al Sayed Al-Bably, and told me to call him at home. Things were fine, not to worry, Hakki assured me as he hung up. But worried I was.

Four months earlier Iraq had invaded Kuwait, and now some 500,000 American and allied troops were in the region preparing for war. In the midst of this I had accepted an invitation to visit this tiny, paramecium-shaped island in the Gulf and explore its tourism potential.

Bahrain, about one-sixth the size of Yosemite National Park, is really an archipelago in the lower end of the Persian, excuse me, I mean the Arabian Gulf (Sunni-ruled Bahrain, because of theological and political differences with Shiite Iran, insists on calling the surrounding waters the Arabian Gulf. Iraq, for the same reasons, does as well)

Left. Bahraini woman dressed in the gauzy black silks of the traditional *abbaya*.

with an inexact number of islands, as the count is ever-changing. By dredging the shallow seabed, new islands are always on the rise, created as seen fit. Bahrain has been a port of call on sea routes between Mesopotamia and India for over 20,000 years, and had flown the flags of a dozen nations until its independence from Britain in 1971. Today it is the most cosmopolitan nation in the Gulf region, where women often hold jobs and drive cars, and men drink freely in swank bars. It is a sophisticated, liberal oasis in an increasingly conservative desert.

I got hold of Al Sayed at home. He apologized for the inconvenience, explaining that the Prime Minister's plane had arrived from Cleveland moments before my Cathay Pacific flight, and that all attention had been diverted to matters of state. He would arrange for the visa and pick me up in an hour.

It was midnight at the oasis when Sayed fetched me. In minutes we were speeding down the causeway connecting the island of Muharraq to the main island of Bahrain. In Manama, a fantasy city built on the sands by oil money, we wound through quiet, dark, curbed and guttered streets. One structure stood out: a squat building lined with colored lights. An Arabic song could be heard, the sound of men chanting like birds of prey. A marriage celebration, said Al Sayed.

We stopped at a gas station, and Sayed pointed out that gas was twice as expensive as in the U.S. Then he took me to the Diplomat Hotel, but it was full of Kuwaiti refugees and American militia. No available rooms. Finally we found a room at the Hilton, and Sayed bade me goodbye, saying that since it was now Friday, the holy day in Islamic brotherhoods, I wouldn't hear from anybody. I should just relax. Not to worry.

I spent Friday worrying and wondering how I got myself into this predicament. Trying to get into the mood, I ate a pita-bread sandwich filled with spiced kebabs at the Al Wasmeyyah coffee shop among dozens of soldiers in unwrinkled chocolate-chip-colored desert camouflage pants and T-shirts that read variously DESERT DEFENDER and DESERT

Below. Sunset at the Tree of Life (Shajarat Al-Hiya), a 400-year-old acacia perched on a hillock in the middle of Bahrain, and the only visible piece of life in all directions.

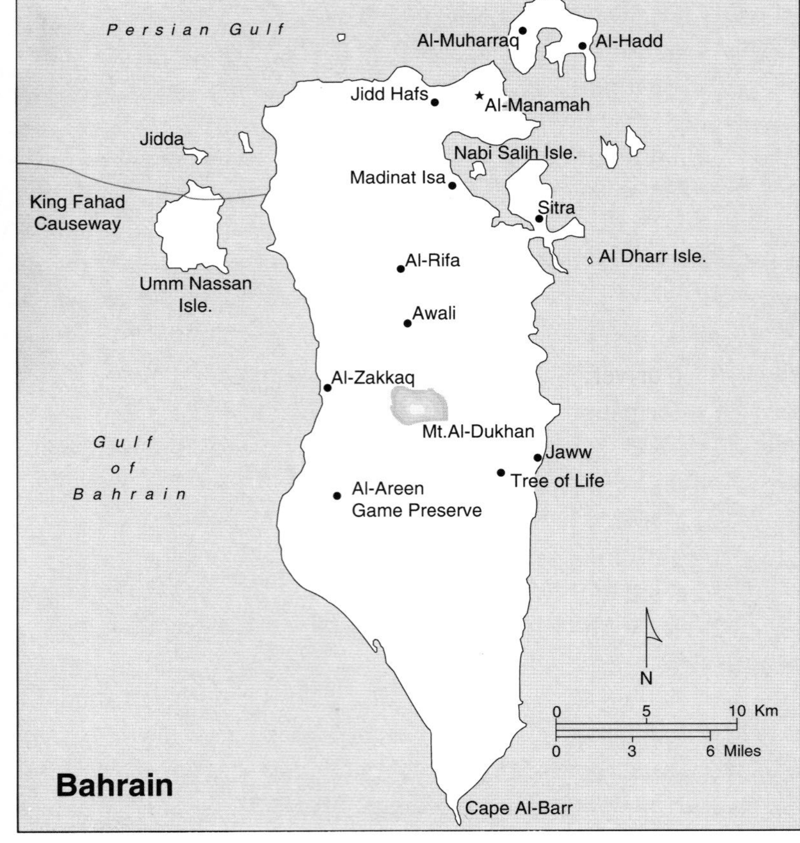

SHIELD, GULF CRISIS 1990. Friday night I wandered into the Hilton disco, and there I stared from the sidelines at a group of Arabs dressed like biblical figures shaking their booties to a Filipino pop group performing Madonna's "Like a Prayer."

Saturday morning I awoke predawn to the harsh amplified ululation of the first prayer call. I wandered outside to a cool, arid wind and the soft blue of a desert dawn. Beyond the hotel walls I could see a series of construction cranes, the national birds of Bahrain. When I returned I got a call from Sheikh Rashid Bin Khalifa Al-Khalifa, nephew of the emir, son-in-law to the prime minister, and the Undersecretary for Tourism, who warmly welcomed me to Bahrain. He would be over in an hour to pick me up. This, I knew, was an honor, and I rifled through my bag to find a fresh shirt. The Al-Khalifa family has ruled Bahrain for over two hundred years. An offshoot of the great Anaiza tribe of northern Arabia, they are related to the Al-Sabah family of Kuwait. Sheikh Ahmed Al-Khalifa, "The Conquerer," drove out the Persian garrison controlling Bahrain in 1782, establishing the dynasty that continues today.

On schedule a shiny new BMW 750i pulled up, and out jumped a boyish, enthusiastic sheikh. He wore a long, white cotton *thobe,* bleached and pressed to a dazzling crispness, and a red-checked *ghutra* headdress held in place by the black *agal* headband, the wool cord once used by Bedouins to hobble their camels. The thirty-eight-year-old sheikh of the Al-Khalifa family drove me to his office on Al-Khalifa Avenue where we were served cardamon-flavored coffee poured from a brass urn into small, handleless cups. Sheikh Rashid explained that the oil wells in his emirate are predicted to run dry in a few years, the natural gas reserves not long after, and the country was looking to diversity, to establish non-oil-based industries. Offshore banking was being considered as a viable alternative, and from the mid-1970s to the early 1980s more than one hundred banks and financial institutions had opened doors on the island. But confidence in the island's future as a safe money haven had eroded in recent months, and business was down 75%. Tourism was something the royal family believed had potential. With money services they had hoped to become the Hong Kong of the Middle East, but there was the fear that the shifting sands of Middle East finance had shifted elsewhere; now they envisioned becoming the Bali of the Gulf. Rashid asked if I could take a look, and offer some thoughts and suggestions. I said of course, that I was keen to see the sheikhdom, and gave him a list of sites I hoped to visit. At the bottom I asked about camping and perhaps climbing the island's highest peak. Not to worry, he would arrange everything.

That one assurance was the last I would need. For the next week things fell into place more neatly than could ever be expected.

The sheikh introduced me to my travelling companions for the next several days: twenty-six-year-old Fareed Hassan, a green-eyed guide; Abbas, our olive-skinned, tie-wearing driver; and white-turbaned and -robed Hakkim, a liaison officer from the Ministry of Information, on board to be sure I didn't somehow breach national security. We started off with a bong. Immediately after leaving the sheikh, Fareed took me into the ocher-colored world of the Middle Ages, somewhere deep in the labyrinth of the market, to a traditional coffee shop, where he ordered up some tobacco packed in a *gidu* (hookah or water pipe). In a room dense with smoke we sat back and puffed until my head spun. At last I staggered to the van and Abbas began driving to the middle of the island to visit the Tree of Life. While whizzing down the perfectly flat highway in our air-conditioned Mitsubishi L300 van with the power roll-back bubbletop, we ran into a traffic jam. The emir's camels were crossing the road. Then, plainly visible on both sides of the highway, the great age of the world seemed to be revealed with sudden poignancy: we passed the famous burial mounds, the largest prehistoric cemetery in the world, a virtual city of dead people dating back to 3000 B.C. Over 150,000 people were buried here in a vast field of

mudpie-like mounds called *tumuli,* stretching as far as the eye could see. Here men had wandered for thousands of years, origins and ends unknown. The dead lie thicker than the living among these hills. Finally, just at sunset, we reached the Tree of Life, a 400-year-old acacia perched on an abrupt hillock, the only visible piece of life in all directions. It was impressive, thirty feet high, with branches spreading almost seventy-five feet across, though on closer inspection a bit disappointing to see trash strewn around its base, the result of visiting journalists. Hakkim said this was the most popular site for foreign news-people, and that he had trekked to this remote site so many times he was tempted to uproot the old tree and move it closer to the capital.

The next morning we drove to the Marina Club on the edge of Manama, a city whose name means "the place of sleeping." There we boarded a sort of super-dhow, an elegant 36-foot-long diesel-powered barge modeled after a Portuguese caravel. The boat was named *Shamsan* ("Two Suns"), and built for the government-owned Tourism Projects Company. We chugged for ninety minutes, dolphins leaping at our sides, black-headed gulls above our stern, on out through a dredged channel, past several U.S. warships, to . . . the Falkland Islands. They were actually a neat set of three man-made islets called Al Dharr, created as a tourism project around the time of the Argentine-British conflict, but somehow the motes took on the timely nickname. A story was making the rounds that a fisherman hit a rock in a storm recently, and started to sink. He issued a Mayday signal on the radio and picked up a British frigate somewhere nearby in the Gulf. When the British officer asked for the location of the accident, the fisherman used the Falkland Islands nickname; the Brit said he was sorry, but he couldn't help with such a distant problem, and hung up.

Vasco da Gama. In 1498 the Portuguese explorer Vasco da Gama reached India via the Cape, opening the whole region, including the Gulf, to European travel.

A gentle sea breeze stirred the tiny islands. Not much bigger than breadbaskets in the turqouise shallows of the Gulf (each was actually about the size of a baseball diamond), they were dotted with umbrellas made with date palm fronds, a small snack bar, which wasn't yet open, and a fort-like tower, which was the bathroom. About a dozen off-duty allied soldiers were lounging on the beach, or in the warm, clear water, while a barbecue was being prepared by the crew of their dhow. Beyond the shimmering blue-green sea was a skyline of refineries and a desalination plant in one direction, and U.S. Navy warships in the other. Not exactly Bali.

Long before the Gulf Crisis, Bahrain had agreed to host the U.S. Navy's Middle East Force, the only U.S. military facility anywhere in the Gulf. It was just northeast of here that the guided-missile frigate *Stark* was hit by an Iraqi missile in 1987, and then repaired in Bahrain's dry dock. Still, the Falklands had promise, something like Green Island off the coast of Cairns in Australia, a day picnic trip to a desert island featuring swimming, snacks, and snorkeling among the seventy-three varieties of coral and just as many tropical fish. Whales, rare green sea turtles, and hawksbill turtles also plied these aquamarine waters, along with nine species of sea snakes, some of them lethal. The waters near these islands are also the favorite habitat of the gentle dugong, or sea cow. This rare, walrus-sized mammal, a first cousin of the American manatee, is so shy and whimsical that ancient sailors often mistook it for a mermaid as it rose from the depths wearing a wig of seagrass. A 1986 aerial survey off the Bahrain coast discovered the largest single concentration of the animals in the world—a herd numbering almost seven hundred.

Back on the main island we visited a dhow boatyard, to see how the real thing was constructed, hand-hewn from teak imported from India. Dhow building peaked in the early part of this century, when the pearl fleets, often more than 5,000 boats, would set sail for the oyster beds in June and return in October. Business dropped radically with the collapse of the industry in the thirties, but dhows were still purchased as fishing ves-

Opposite, top. A traditional wooden fishing craft, or *shu'i*, powered by diesel engines, is the chief vessel of Bahrain's commercial fishing fleet centered at Sitra.
Bottom left. Since Shariah (or Islamic) law does not apply in Bahrain, veils are more of a social institution than legal dictate.
Bottom right. The Arabian thoroughbred horse was once used as a means of transport and a mount in desert battle, but today is kept for its sheer beauty and sporting qualities.

Dugong.

sels and for trade and transport to Saudi Arabia. The latter market disintegrated when the 1.2 billion-dollar, 15-mile-long King Fahd Causeway opened in November 1986, the largest single construction job in the history of the Gulf, and the most expensive bridge in the world. It stretches from the mainland like an arm reaching for a pearl, and it's now an hour drive from the crude Saudi city of Dhahran to the bars and beaches of Bahrain. Many of the old boatbuilders have lowered their sights and now make a living building models for sale to tourists. In a room near the boatyard I found a craftsman sawing away at a tiny lateen sail for a model dhow. Just as in crafting the big versions, the artist used no plans or power tools, and he had memorized all the measurements and proportions. For inspiration, on his wall there were a number of posters, of Rambo, of Arnold Schwarzenegger in *Commando,* and one poster showing a stealth fighter leaving Langley Air Force base for the fifteen-hour flight to the Gulf, with the legend: "In Slickness and in Stealth."

The next morning I awoke early again and went out to run along a strip of reclaimed land parallel to King Faisal Highway. It was a strange run in a pearly mist. At a roundabout I circled around a statue of a giant pearl held aloft by cement fingers. On one side of the monument was a row of brushed-steel skyscrapers; on the other a coral-encrusted coast with a row of ibis feeding in the tidal marshes. Along the route I passed several U.S. servicemen also running, and in the distance a black penguin seemed to be dashing toward me. As it approached I saw it was a lone woman jogging along in gauzy black silks (the traditional *abbaya*). Was she young or old, ugly or beautiful? All I could see of her were her henna-tattooed hands, a picture of Arab probity running for reasons I would never know.

Fareed was ever cheery, and I asked him how this was so when such a cloud of doom hung over the region. He said Bahrain, perhaps the most vulnerable nation in the Middle East, had come to accept insecurity as a foundation of life, and that gave people strength and confidence in the future. Life was pretty good on his little island, he explained. With just over half a million people there was no significant air or water pollution, no homeless problem, no severe economic problems. The monarchy has the fourth-highest per capita GDP in the Arab world, after Qatar, United Arab Emirates, and Kuwait. It is a place Fareed would never leave, even if occupied by Iraq, something that had happened once before, when the Assyrians (another aggressive Iraqi tribe) controlled the island from around 1200 B.C. to 50 B.C. Even if bombed with chemicals, he claimed he could tape his windows closed during the attack, and after twenty hours the air would be clean. His family was his sanctuary, his home inviolable. Fareed seemed the living embodiment of the legendary inhabitants of Dilmun, people who lived in an earthly paradise of unimaginable beauty and plenty, where only to exist was to experience happiness.

Below. At the royal racetrack a dark bay belonging to the ruling Al-Khalifa family is prepared for the weekly race. *Opposite.* Date palms at the Virgin's Pool, a large freshwater spring which traditionally supplied much of Bahrain's sweet water.

Above. The Arad Fort is an Arab defense structure possibly built during the latter half of the sixteenth century.
Far left. Strolling through the old quarter of Muharraq past nineteenth-century pearl merchant's homes. A minaret, where the call to prayer is made, is in the background.
Left. A Bahraini traffic jam—the emir's camels crossing the highway.

That evening, as we were driving past a giant Alice-In-Wonderland teapot built in front of the Al Fateh Grand Mosque, the sun began a fiery descent into the sea across the bay. I asked Abbas to stop the van for a photo, and we pulled up in front of an old wooden dhow sailing to shore after a day of fishing. I pointed my camera as the captain bailed water from its bilge with a wooden bucket, and was thrilled at the prospect of photographing this remnant of an earlier time. But I couldn't compose the shot as I hoped, as though taken in another century. In the background were rows of polished skyscrapers of marble and glass, and the peaks of Manama. As the sun made its final bow, across the stern of the dhow a windsurfer sailed into the viewfinder. For a moment I could focus simultaneously on past and present, as if they were stereoscopically in a single image.

An hour later the sheikh picked me up in his new Mercedes-Benz and took me to the Bahrain Arts Society, of which he is president. Along the way he puffed on his cigar and popped a cassette into the Blaupunkt. It was Milli Vanilli, singing "Blame It on the Rain," and I mentioned that recently the group had been exposed as not singing the lyrics of their songs. The sheikh smiled and said he knew, and figured the title of the song had always been "Blame It on Bahrain." The sheikh was in a buoyant mood. It turned out art was the sheikh's true passion. Not only was he a great patron of the arts, but he was an artist as well, a talented painter of oils and watercolors. He showed me several of his works about to go up for an exhibit, and then took me to a Tandoori restaurant where we curried flavors with several members of the Arts Society. Over a couple bottles of Lebanese wine there was a brief debate over whether photography is true art. One painter was quite sure it wasn't. "Whether you have an F4 or an F15 or F16, it doesn't matter, because the tool is doing all the work," he stated emphatically.

Mosque in Bahrain, sketched by a nineteenth-century European visitor.

The next day we drove to the southwest corner of the thirty-mile-long island. It was midmorning, and the farther south we travelled, the more inhospitable and colorless the landscape became. It was as though Nature had exhausted every tint in her paintbox. Then, like a mirage, a scrim of green wavered into view: the remarkable Al Areen Game Preserve. A personal project of the Crown Prince, the park is an experiment in conservation, sheltering indigenous Arabian animals threatened with extinction, as well as several North African species. Established in 1975, it has been stocked with over 100,000 plants and trees, 10,000 varieties of birds, and more than 500 different animals. Among these are the addax, impala, Persian gazelle, fallow deer, Grant's zebra, springbok, sable antelope, eland, zebra, waterbuck, ostrich, wildebeest, topi, wild sheep, and most significantly, the scimitar-horned Arabian oryx, extinct in the wild, with only 400 surviving in captivity. This is the first project of this type in the Middle East, and the most successful. The Kuwait Zoo was also renowned for its collection of area wildlife, but in the wake of the Iraqi invasion, and the Allied food embargo, all the edible animals were eaten by soldiers.

Al Areen seemed positively lush, and gave a sense of what much of Bahrain must have looked like before overgrazing by camels and goats denuded much of the natural vegetation. I was impressed with the preservation efforts. Still, as we wandered around the grounds there was a reminder that Bahrain itself was a delicate and crushable flower, that total ecological disaster could be just around the corner: the roar of jet engines cut the air, and scores of antlered heads turned skywards. I knew the sound, and instead trained my eyes on a poster from the Environmental Protection Committee of Bahrain. It showed a dolphin crying to a man tossing trash into the seas, "Stop—You are killing me!!"

On the way back north to the Oz-like capital, Manama, we trailed several expensive cars displaying the bumper sticker KUWAIT FOR KUWAITIES. Then about halfway back Hakkim told Abbas to pull over to a dried-mud hut. Hakkim wanted to visit his friend, Al Hagry Ajab, a sixty-five-year-old master in the ancient art of falconry. Originally falcons were trained to catch desert game for food. For centuries small gazelle, hare,

Opposite, top. Inside the Grand Mosque on the Manama corniche. The country's newest mosque showcases $23 million of Carrara marble imported from Italy.
Bottom. Young antelope at the Al Areen Game Preserve, a personal project of the Crown Prince.

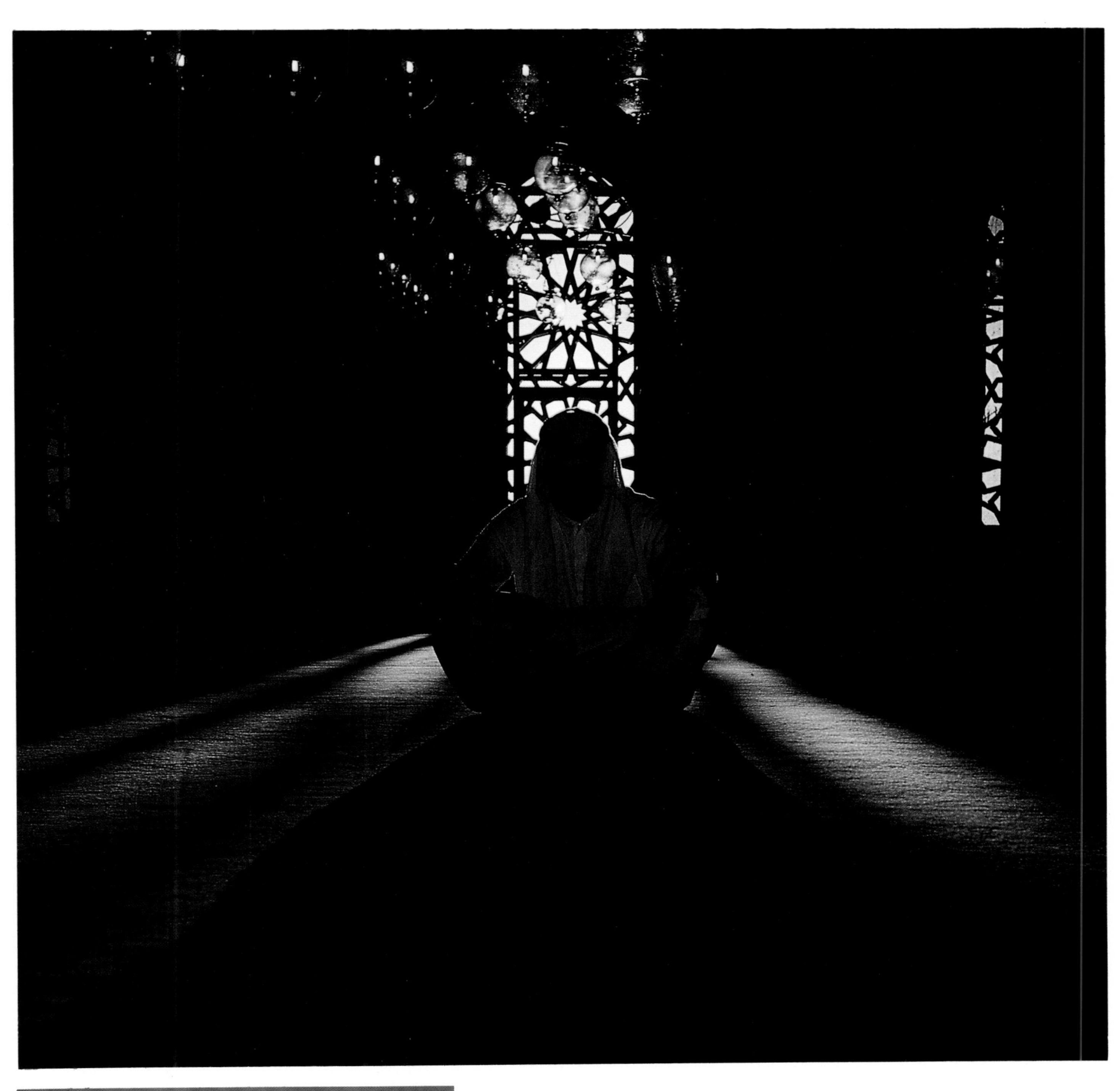

and bustards were caught, but with the reduced dependence on the desert as a source of food, falconry became a sport. Mr. Ajab's face was creased and leathery from the sun, his eyes sharp and dark, like those of a falcon, with a beak even more so. He was happy for the company, and brought out one of his prize saker falcons on a leather cuff on his left wrist. He had been raising and training falcons for forty years, and he sold them for about 4,000 dinar ($12,000) to the royal family and other wealthy sportsmen. He told us there were over one thousand serious falconers in Bahrain. The buyers travel throughout the Middle East for competitions, often flying first class and insisting the falcons fly with them in the adjacent seat.

Back in Manama we took a tour of the *souq*, the bazaar. Between Pizza Hut and Hardees and some ultramodern air-conditioned malls there were relics of a life before petro-dollars, such as the shop belonging to Hassan M. Al-Arrayed, a gap-toothed seventy-two-year-old pearl merchant. His father, blind from birth, was one of Bahrain's premiere pearl merchants, and his father's father used to don tortoise-shell nose clips and tie stones to his feet in order to dive to depths of eighty feet without air tanks, wet suit, or goggles, to fetch the "fisheyes of Dilmun." Sumerian legend has it that a king named Gilgamesh, the hero of the world's first epic odyssey, learned that a sort of fountain of youth existed on Dilmun. Once he sailed to Dilmun, he was told to dive for the flower of immortality at the bottom of the sea. Attaching stones to his feet he dove, and brought the flower—a fantastic pearl—back to the surface. But while Gilgamesh took a bath a serpent swallowed the pearl, whereupon it shed its skin and emerged new and shining and youthful. The possibility of eternal youth for mankind was lost forever.

Seals from the Dilmun period were a means of property identification when writing was known only to a privileged few. Many of them bear intricate designs of gods, heroes, and animals.

Al-Arrayed started in the business when he was ten, and has been in the same stall since. From an old safe behind his desk he retrieved a ragged, red felt handkerchief and poured out over $100,000 worth of brilliant natural pearls. The oyster beds around Bahrain were justly famous for thousands of years for the unique pearls they produced, pearls with a strange, lustrous sheen, attributed to the rinsing received from fresh water that jets from the ocean bottom. The Kanoo family, one of the richest families in the world, rose to prominence as pearl traders one hundred years ago. Now every year they supply some four thousand ships in the Gulf with essential services, such as docking and cargo handling, and the family head is vice-chairman of Bahrain-based Investcorp, which now owns Tiffany & Company and Saks Fifth Avenue.

I picked out a single, grape-sized pearl the color of hand cream, with a light pink emanating from its center, and asked Al-Arrayed the price. I had picked a good one, he said, for it sold for 5,000 dinar ($15,000). The pearl industry had ground to a near-halt in the 1930s, the result of competition from the newly established Japanese cultured pearl industry, and decline in the yield of the Bahrain pearl banks, fished out after 5,000 years. Where once there were more than five hundred traders in the *souq al lulu* (pearl market), now just a few survived. And their shops are replenished with less than twenty-five pounds of new pearls found each year, most from amateur divers who rarely plunge beyond a depth of ten feet. Nonetheless, Hassan M. Al-Arrayed did not seem to be suffering too dearly. Despite his tiny, shabby shop, Mr. Al-Arrayed proudly told me he had two new Mercedeses parked in his driveway at home.

On the way back we stopped in Muharraq and met Yasser Yousif Al-Ansari, the last Arabic coffeepot maker, who had been plying his trade for ten years. He polished the spout of his latest work, a shiny brass-and-copper model, and then held it in the sunlight flooding in from the street. As I admired his artistry he glared from beneath his crocheted cap, and finally spoke in an angry timbre. He said he couldn't make enough to make ends meet practicing his ancient craft. It took him ten days to produce one coffeepot, and it often took six months to sell one. So he had decided to work at the shipyard, and abandon

Opposite, top. Camping, Bahraini style. A colossal Bedouin tent, lined with Persian rugs and perfumed by incense, set up in the central desert for a night of "roughing it."
Bottom left. Originally falcons were trained to catch desert game for food, but with reduced dependence on the desert as a source of food, falconry became a sport.
Bottom right. Serious falconers travel the Middle East for competitions, often flying first class and insisting their falcons fly with them in the adjacent seat.

Arabian dance.

Arabian concert.

Below. A traditional Arab dhow, hand-hewn from teak imported from India. *Opposite.* The mysterious woman inside the luxurious camping tent is photographer Pamela Roberson, who finally surrendered to the culture and went native.

his beloved craft. In the modern one-crop economy of Arabia, he was lost in another time.

Next we visited Ain Adhari, the Virgin's Pool. Like all the sweet water on Bahrain, the pool is fed from an underground channel that has welled up through fissures in the limestone foundations, discharging water accumulated far away under the landmass of Arabia. (Bahrain in Arabic means "two seas," likely referring to the salt water of the Gulf and the sweet water below the island. Water, of course, is more important than oil on the arid sands of Bahrain, and five thousand years ago the inhabitants worshiped Enki, the god of sweet waters under the earth. Today the reverence is directed towards the desalination plants, which have turned the island into the highest per-capita water consumer in the world—225 gallons a day, compared with the international average of 57.) Legend has it a group of virgin girls were walking in the desert when a horseman galloped over and hijacked the most beautiful. The others begged for help from the heavens, and the ground suddenly split open, whereupon all the girls fell in and became colorful fish. I could see no fish in the Virgin's Pool, just dirty water. And as I gazed a man in a lycra biking outfit wheeled up on his ten-speed, stripped to his shorts, and jumped in, defiling the chaste reflection of a date palm.

We took a shortcut home, across the industrial island of Sitra, and out the right window we could see a U.S. destroyer anchored. Hakkim's fingers constantly twirled a set of amber worry beads. Then he turned his head to the other side where an ass was blocking traffic on the road. "Look," he said, as an uncharacteristic smile spread over his face, "Saddam Hussein."

Once I was back at the Hilton, Sheikh Rashid called to say a car would be by in thirty minutes to take me camping. I asked if I should pack my sleeping bag, my Thermarest, my Sierra Club cup, my flashlight. He laughed and said no. Not to worry. Everything had been arranged.

Not completely assured, I pinched a blanket and sheet from the Hilton and walked out front. There I met Sheikh Khalifa Abdulla Mohd Al-Khalifa, superintendent for Tourism Relations and another nephew to Sheikh Isa Bin Sulman Al-Khalifa, the tenth emir of the Al-Khalifa family. Sheikh Khalifa had a pleasant hint of the Tigris and Euphrates in the bones of his face, and he smiled broadly as he stuffed my kit into the trunk of his Mercedes 500 SEL. Soon we were gliding through a thick fog, one that covered us like a shroud, towards a favorite camping site.

About forty-five minutes later we pulled off the highway onto a dirt road, and then off the dirt road over the broken, monochromatic desert towards a small hotel . . . but no, as we drove closer I saw it was not a small hotel, but in fact a large tent. A colossal Bedouin tent. Workers were putting on the finishing touches as we parked the car, and I wandered inside to an astonishing sight. The entire interior of the tent was lined with inch-thick Persian rugs, padded with gold-embroidered cushions, and perfumed by incense. A partition created a separate section where I found my bed, a comfortable affair complete with fluffy pillow, crisp sheets, and an Arabian horse-hair blanket. Electric lights lit the entire structure. Outside was a separate tent, also well-lit, which was the toilet. While I was admiring the place a van pulled up, and out popped several workers who extracted a table and chairs and positioned them in the tent. A cook started up the stove on the tailgate, while three tuxedoed waiters appeared and began to set the table with bone china and crystal. One served me Arabian coffee (boiled cardamon with some cloves and a bit of Nescafe) from an elegantly shaped and intricately patterned silver *dalla.* Thirty minutes later I was sitting down to an exquisite banquet, featuring *hamour* (grouper), giant prawns, *houmos* (garbanzo bean paste), *baba ghannouj* (smoked eggplant with sesame cream), *dolmas* (grape leaves stuffed with rice and ground lamb), *kibby* (cracked wheat with ground lamb and pine nuts), *shish tawook* (breast of chicken marinated in olive oil), tabouleh salad, cucumber yoghurt, and bowls of dates, pomegranates, and figs. The tuxedoed waiters hovered over every course and kept our glasses full of Emirates, the favorite bottled spring water. Camping would never again be the same.

Just as the *baklava* smothered with pistachios and honey sauce was being served there was a call for me on the car phone. It was Sheikh Rashid, asking if I wanted to join him for a special performance of *Phantom of the Opera* being presented at the Arad Fort. I said certainly, and was whisked to his plush couch in the front row, where I was served sweet tea in glass cups as we watched while Carlotta sang and the Phantom made his way among the parapets. I was back at my tent by ten o'clock. Before retiring I looked out to constellations I didn't recognize, until I realized they weren't stars at all, but the twinkling lights of an oil refinery.

Geometric wall hanging.

I awoke to the sound of wild saluki dogs barking at the dawn. Outside the air was soft and moist; a glimmering haze hung over the landscape. I changed into my shorts for a run. I took off down an old road, and my nostrils filled with the foul stench of natural gas as I ran along a series of pipelines and up a small rise. There I came to a ditch surrounded with red and white markers, the national colors of Bahrain. A plaque at its side proclaimed that this site, Jebel Al-Dukhan, was Oil Well No. 1, where the black gold was first discovered in the Middle East on June 1, 1932, flowing from a depth of 2,008 feet. At the time Bahrain had no paved roads, electricity, flush toilets, or radios. Two years later Bahrain became the first Arab country to start exporting oil. The fruit had been bitten, the box opened.

I was also, I realized, at the foot of the Great Smokey Mountain, the highest peak in Bahrain, pressing its jagged yellow teeth into the morning sky. Now was my chance to climb the Everest of Bahrain. I began scrambling up a limestone slope whose sides broke off like crumbs of a biscuit. I huffed and panted, fell and scraped a knee, and stopped to catch my breath several times. At last I crawled to the summit, and in a spanking breeze I stood and raised my arms in a Rocky salute. Top of the world, Ma. I was 451 feet high.

As I stood on this single pillar of wisdom, watching the horizon fall away at every direction, overlooking a barren but rich landscape, I admired its dizzying beauty. The wind died, and it was utterly still, the only movement the lazy turning of my own thoughts. Then an F-16 flew over my head and broke the sound barrier. This was the end of innocence, once again.

Madagascar

Forest in a Sack

I would rather die than eat anything taboo.

One tree does not a forest make.
—Malagasy proverbs

It is *fady* to kill a lemur. The punishment is ill health, and five years in jail.

Madagascar, the planet's fourth-largest island, floats 250 miles off the east coast of Mozambique in the southwest Indian Ocean. The Afro-Indonesian people there have long governed their lives with a series of social taboos, or *fadies.* And a long-time *fady,* rooted in the commands of the *razana,* the Ancestors, was that it was wrong to kill the little button-eyed primates called lemurs. These ancient relatives of monkeys, apes, and humans are found only on this island, which apparently rafted away from the vast bulk of the African continent 165 million years ago. Yet even today, in a world of heightened environmental consciousness and recognition of the accelerating loss of species, lemurs are still being killed; sometimes to be served at the tables of wealthy foreigners who will pay a little extra to have a taste of the exotic. In the 1990 Marlon Brando film *The Freshman,* the plot revolves around a moveable restaurant that serves endangered species to high-rolling epicureans. In a case of life imitating art, I heard a rumor that a restaurant existed in the Malagasy Republic that served lemur.

October is the burning season for Madagascar. When I approached Antananarivo, the 200-year-old capital, on an Air Mauritius Boeing 737, the air was thick with smoke, the landscape parched and coughing. As subsistence farmers below were clearing crop and pasture land and scorching trees to create charcoal I struggled to fill out the customs and immigration form on my lap. After twenty-five hours of flying from San Francisco it was not a simple chore on the coarse, brown customs form that seemed to be made of cheap toilet paper.

While waiting for the baggage in the Ivato International Airport I visited the men's room, and discovered in a world of disappearing species there was yet another. The attendant offered to sell me toilet paper, as there was none in the stall. "There is a shortage in Madagascar," he explained, and I knew why—it was being used for customs forms.

Within an hour I was on another Air Madagascar (a.k.a. "Air Mad") flight, to a large island off the northwestern shore of the Mozambique Channel, Nosy Be ("Big Island" in Malagasy). From the air there was a muscular poetry to the brown, bare landscape, with raw red rivers, like broken veins, bleeding to the sea. A Soviet cosmonaut said

Opposite. A black lemur of Nosy Komba. Diurnal and fruit-eating, these creatures are remarkably trusting of people.

Below. A lepilemur plays hide and seek.
Opposite, top. Perinet guide Laurette Nirina and a Parson's chameleon.
Middle. The Breakfast Club.
Bottom. A saying on Nosy Komba is "If you feed the black lemurs you will be rich; if you kill one, you will die."

the Texas-sized island was the only landmass he could identify from space, because it was surrounded by a halo of rust-red sea, the color of the lateritic topsoil relentlessly scrubbed off its denuded surface by wind and rain. The microcontinent of Madagascar is the most eroded place on earth. Man has grievously wounded this estate. Some estimate 90% of the Great Red Island's forests have been destroyed, and that it continues to lose 375,000 acres a year, a rate that will insure a totally bald island within a lifetime. The impoverished peasants of preindustrial Madagascar deliberately torch the rain forests for fuel and agricultural land. After a few seeding seasons the thin soil is depleted, erosion sets in, the tired, overtaxed land is abandoned, and a new wedge of forest is obliterated, a dangerous cycle that promises to turn the land that time forgot into a dead zone. In the fifteenth century Arab traders called Madagascar the "Isle of the Moon." As the twenty-first century approaches it seems their epithet may have been prophetic. The avoidance of a land with the life and scape of the moon, the protection of the remaining woodlands, is a race against starvation, ignorance, and time.

As we began our descent, bright Ricky Nelson and Roy Orbison tunes playing over the loudspeaker contradicted the scene below. But as the wheels lowered, the landscape turned green and a flock of snowy egrets fluttered from the parasol of a gigantic glossy frond. To further abet the mood change as I stepped off the plane a lei of fresh frangipani and bougainvillea was placed about my neck by a smiling Malagasy girl. She led me outside to a row of stands where plump giggling women were selling stacks of vanilla, bottles of mango, and peppers in vinegar. This suddenly seemed a happy place. I climbed into a red candy-striped two-cylinder Citroen Deux-Chevaux with a $100 bill stuck on the front windshield. Looking closer I saw a profile of Madonna where Ben Franklin belonged, and in place of the nation's name were the words "Altered States of

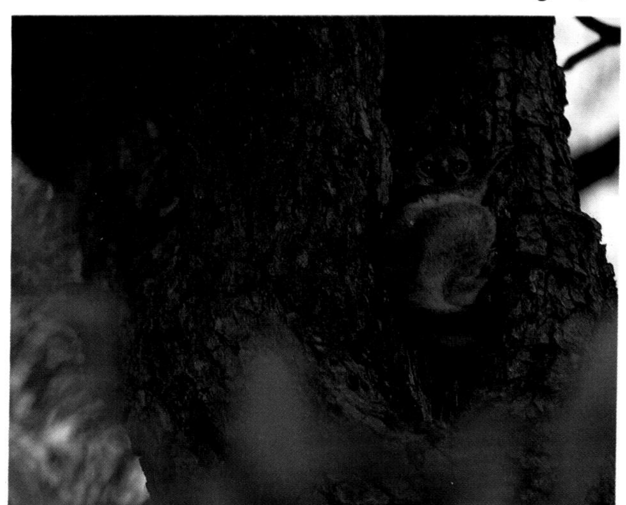

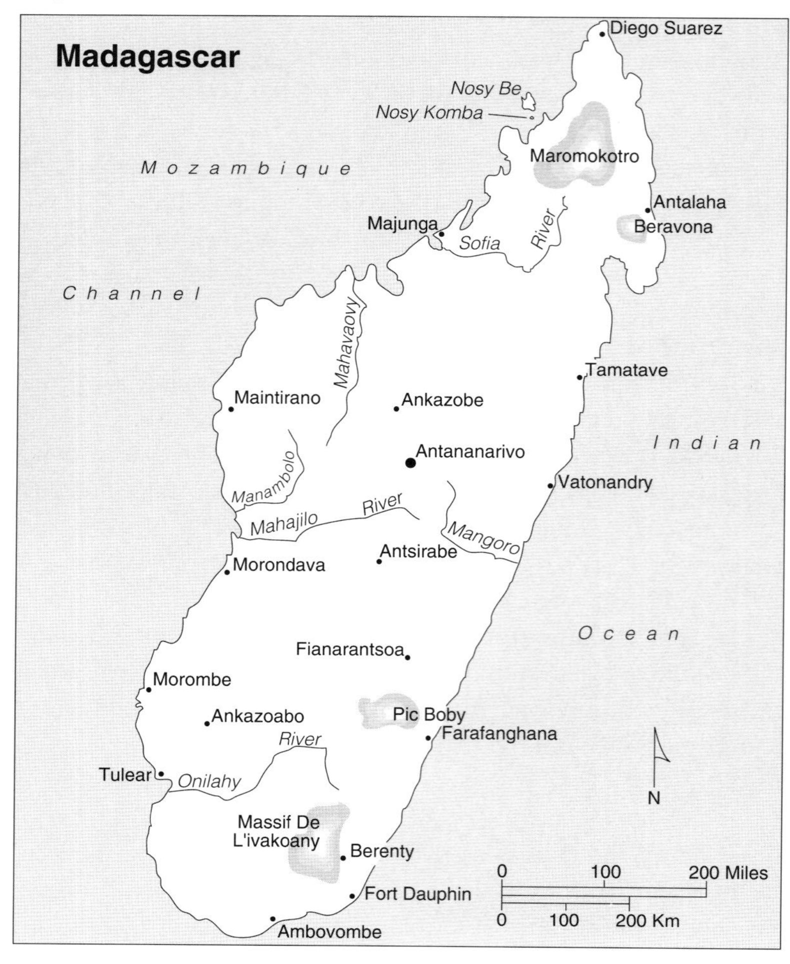

Madonna." We bumped down a sunroasted road trying in vain to avoid the potholes that are an unstudied endemic species in Madagascar. The edges of the road were lined with kapok and pollarded yellow-flowering *ylang-ylang* trees from which a perfume essence is extracted. We rolled past vast sugar cane fields and balloon-shaped concrete huts built in 1921 as cyclone shelters. I checked into the Les Cocotiers hotel, and on the wall was the requisite local art, a sisal fiber tapestry of a village scene, and next to it a lizard I would get to know. He hung on the wall like bad art. I named him Art Gecko.

At lunch I spied another *vazaha,* a paleface, and wandered over to join eighty-year-old George Rosemond, a retired surgeon from Philadelphia. In a country that has only allowed tourists since 1984, and which last year hosted fewer than 400 American visitors, I was glad to find another homeboy. We each ordered the local special, a pork stew with sticky rice spiced with green leaves from a flower called *anamalaho,* which left tongues stinging as though stuck in an electric socket. This led us to a couple of bottles of Eau Vive, a French bottled water. But we soon discovered that beer in Madagascar is cheaper than bottled water, so we switched to hearty Three Horses beer, into which George poured his own airline-sized bottles of Canadian whiskey. Several bottles of Three Horses led George to tell me of his ordeal in getting to Madagascar. A week earlier he had made the same grueling flight I had to Mauritius from the United States, but when he caught his connecting flight to Madagascar, he was unaware there was an intermediate stop. When the plane landed he followed the crowd, went through immigration, where his passport was stamped, and proceeded to customs. When his bags never appeared, he went through customs to see if the promised travel agent was there to meet him. But there was nobody. So he flagged a taxi and asked to be taken to the Antananarivo Hilton, as described on his itinerary. But the taxi driver, who only spoke French, shook his head no and left. Bewildered, George sought out an airline employee, who courteously listened to his plight, then smiled and explained that George was unfortunately not in Madagascar as he thought. He had stepped off the plane one stop too soon and was on the French department (district) of La Reunion Island.

Pirate with maiden.

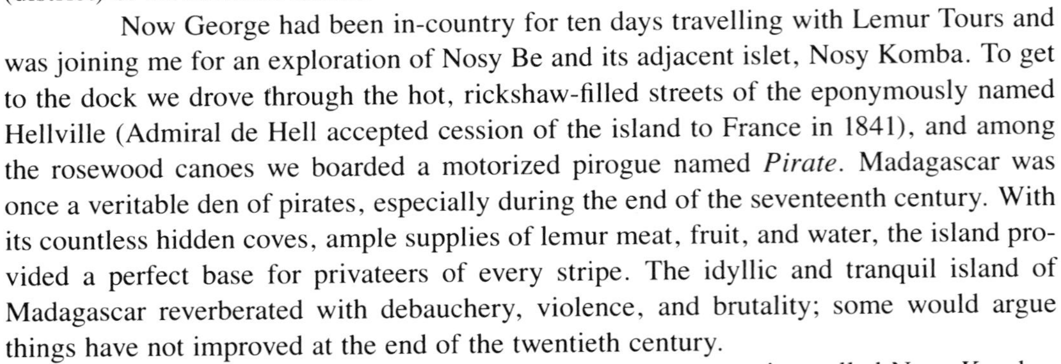

Now George had been in-country for ten days travelling with Lemur Tours and was joining me for an exploration of Nosy Be and its adjacent islet, Nosy Komba. To get to the dock we drove through the hot, rickshaw-filled streets of the eponymously named Hellville (Admiral de Hell accepted cession of the island to France in 1841), and among the rosewood canoes we boarded a motorized pirogue named *Pirate*. Madagascar was once a veritable den of pirates, especially during the end of the seventeenth century. With its countless hidden coves, ample supplies of lemur meat, fruit, and water, the island provided a perfect base for privateers of every stripe. The idyllic and tranquil island of Madagascar reverberated with debauchery, violence, and brutality; some would argue things have not improved at the end of the twentieth century.

It was a forty-five-minute sail to the volcanic outcropping called Nosy Komba, and along the way dolphins leapt at our stern as though encouraging our voyage. Once on the busy beach we waded through women snugly wrapped in *lambas* (technicolored cotton shawls) and warm-faced, barefoot children whistling through bougainvillea blossoms, all chatting in a mellow, polysyllabic tongue related to a language in central Borneo. Behind the main village of Ampangorina we stepped past an old blind man strumming a resonating tubular bamboo box with metal strings called a *valiha* and into the black lemur reserve. Scores of these endearing critters lolled in the crotches of trees, swung from rafter-like horizontal branches, bounced lightly through the trees like arboreal kangaroos, and scrambled to human shoulders to beg bananas with big, irresistible, imploring eyes. The males were truly black, sable-furred from head to toe, faces punctuated with wide, inquisitive, lemon-colored eyes. The white-bearded vulpine females sported fashionable golden-furred coats, and earmuffs of white Einstein hair.

Diurnal and fruit-eating, these creatures were remarkably trusting of people. Not a trace of fear fogged their translucent eyes, a result of an isolated evolution without predators since the Age of the Dinosaurs. Conservationists say eighty percent of the flowering plants (some 10,000 species) and ninety percent of the beasts on this singular island, including the thirty-one species and forty subspecies of lemurs, are found solely on Madagascar and its satellite islands. The area is the most outstanding living laboratory of evolution in the world, far more so than the Galapagos. It is, in a way, like Sir Arthur Conan Doyle's Lost World. When the cataclysmic forces of plate tectonics tore the island from the African continent, its cargo of plants and animals merrily evolved on a parallel but separate track from the rest of the Earth's ecology. "It is as if time had suddenly broken its banks and flowed down to the present in a completely different channel," wrote naturalist Alison Jolly in her classic book *A World Like Our Own*. Because of the island's unequalled levels of endemism, razing a patch of irreplaceable primary forest, or poaching an animal, can have more devastating repercussions here than just about anywhere else. Two new primate species, the golden bamboo lemur and the golden-crowned sifaka, have been discovered since 1987, and one thought extinct, the hairy-eared dwarf lemur, was rediscovered in nature's attic in 1989. The rosy periwinkle, the source of a medicine used to treat childhood leukemia, grows here. It's quite possible a cure for AIDS might be in a patch of forest about to be burned.

Where is the trader of
London town?
His gold's on the capstan
His blood's on his gown
And it's up and away for St.
Mary's Bay
Where the liquor is good and
the lasses are gay.
—Seventeenth-century sea chanty about Madagascar's favorite pirate base

While we disported with the black lemurs, Pierrot Rahalijaon, an operations manager for Lemur Tours, told me that the tradition among the Nosy Komba villagers was strong: "If you feed the lemurs you will be rich; if you kill one, you will become sick and die." I asked Pierrot if he had heard of a restaurant that actually served lemurs. As far as he knew no such eatery existed.

Under an old fig tree cabled with lianas we lunched on carangue fish. Not far away a white-necked pied crow waited patiently for our scraps. It seemed like paradise. It was hard to imagine, looking out over the curved, coconut-palmed white beach to water like tourmaline glass, that this was an Eden on the edge, with several species beyond. Some have likened those who hunt lemurs and other antique wildlife to Nazis, and the notion has an eerie sting. In 1934 Hitler devised a plan to deport European Jews to Madagascar, where it was hoped they would all die from tropical diseases. But when the British took the northern part of La Grande Ile from the Vichy government, the plan was abandoned. In its place Hitler sanctioned the Final Solution.

The following day we flew back to Antananarivo, and I hooked up with a guide named Serge, a thirty-five-year-old who had lost his left eye to an arrow playing cowboys and Indians when he was a small boy. Serge's surname was Harizo, one of the shortest in the country (the king who conquered and united the highland clans in 1794 was named Andrianampoinimerinandriantsimitoviaminandriampanjaka). I asked Serge if he knew of the infamous restaurant. He said he had, just recently, but didn't know if it really existed, or where it was, but he promised he would put the word out. In the meantime he would take me to the special reserve at Perinet, sixty miles to the east, halfway to the coast.

Left. The spiny desert in the south of Madagascar features sisal plants and "the tree where man was born," the bowl-bellied baobab.

Left. A scarred crater lake on Mont Passot, the highest peak on Nosy Be.
Above. The soft-toy cuddliness of the Sifaka lemur makes them the most appealing to humans. Unlike other lemurs, Sifakas rarely come down to the ground.

As we bowled down the road that switchbacks off the eucalyptus and pine-lined Hauts-Plateaux of Antananarivo there was a haze both inside and outside the car: outside from the incessant burning, inside from Serge's Dunhill cigarettes. Even though Serge had met less than a score of Americans in his guiding career, he had seen enough to make observations, and one was that to Americans smoking is *fady.* I agreed and asked if, out of cultural reverence, he could put out his smoke.

The trip took us past gneissic slopes covered with coarse grass and lavaka, great fan-shaped erosion gashes in the hillsides, looking like fleshy wounds inflicted by some savage giant. And then there was the burning. In the distance the fires were a necklace of streetlights. Up close the defoliated, barren landscape was much uglier.

Once at Perinet we took a nocturnal walk down the Ancienne Route Nationale Numero Deux into the protected montane forest. The silence was carpet-deep, and our guide, Laurette Nirina, spoke in a cathedral hush as she swept a weak flashlight across the trees. Every few minutes her narrow beam would catch the round, hot-coal eyes of the world's smallest primate, the reclusive *antigi* or mouse lemur, about the size of a newborn kitten, and those of the fat, brown, greater dwarf lemur, as well as the eyeshine of tiny tree frogs and chameleons. (Two-thirds of the world's chameleons slink on this island alone, including the biggest and the tiniest.)

Detail from an eighteenth-century map of Madagascar.

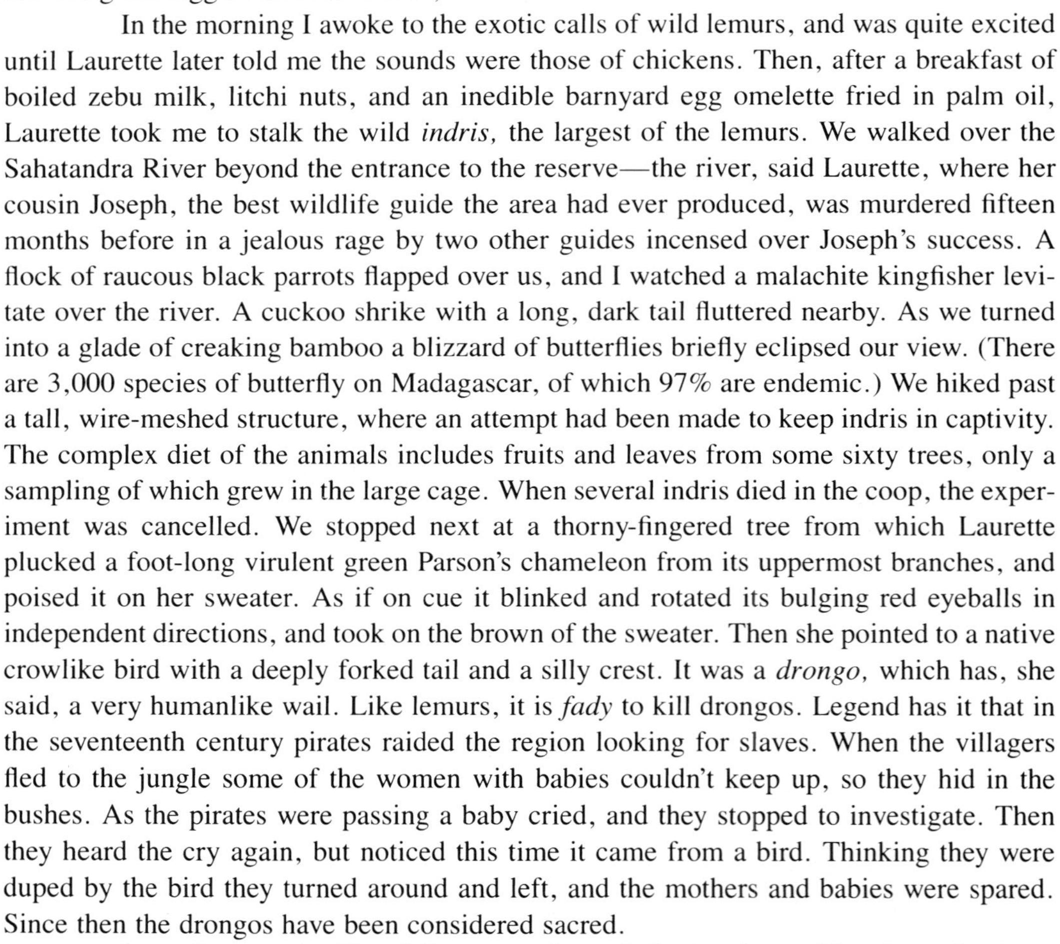

In the morning I awoke to the exotic calls of wild lemurs, and was quite excited until Laurette later told me the sounds were those of chickens. Then, after a breakfast of boiled zebu milk, litchi nuts, and an inedible barnyard egg omelette fried in palm oil, Laurette took me to stalk the wild *indris,* the largest of the lemurs. We walked over the Sahatandra River beyond the entrance to the reserve—the river, said Laurette, where her cousin Joseph, the best wildlife guide the area had ever produced, was murdered fifteen months before in a jealous rage by two other guides incensed over Joseph's success. A flock of raucous black parrots flapped over us, and I watched a malachite kingfisher levitate over the river. A cuckoo shrike with a long, dark tail fluttered nearby. As we turned into a glade of creaking bamboo a blizzard of butterflies briefly eclipsed our view. (There are 3,000 species of butterfly on Madagascar, of which 97% are endemic.) We hiked past a tall, wire-meshed structure, where an attempt had been made to keep indris in captivity. The complex diet of the animals includes fruits and leaves from some sixty trees, only a sampling of which grew in the large cage. When several indris died in the coop, the experiment was cancelled. We stopped next at a thorny-fingered tree from which Laurette plucked a foot-long virulent green Parson's chameleon from its uppermost branches, and poised it on her sweater. As if on cue it blinked and rotated its bulging red eyeballs in independent directions, and took on the brown of the sweater. Then she pointed to a native crowlike bird with a deeply forked tail and a silly crest. It was a *drongo,* which has, she said, a very humanlike wail. Like lemurs, it is *fady* to kill drongos. Legend has it that in the seventeenth century pirates raided the region looking for slaves. When the villagers fled to the jungle some of the women with babies couldn't keep up, so they hid in the bushes. As the pirates were passing a baby cried, and they stopped to investigate. Then they heard the cry again, but noticed this time it came from a bird. Thinking they were duped by the bird they turned around and left, and the mothers and babies were spared. Since then the drongos have been considered sacred.

Laurette moved with a feline grace through the tangle, and I awkwardly stumbled behind, imagining I was playing Sigourney Weaver playing Dian Fossey. After we crawled through the damp, orchid-festooned forest for an hour, the air splintered with the spine-chilling, unearthly call of the indri. The sound grabbed me by the scruff of the neck and pointed my ears skywards. The word "lemur" is from the Latin for ghost, and the aural-linked etymology seemed clear as the calls seemed to echo from the hereafter. They were loud, eerie, and childlike, sounding to me like some kind of sad human saxophone. The ghostly symphony lasted only three minutes, then abruptly stopped, leaving behind the susurrous hum of insects and the flatulent honks of frogs, feeble by comparison. Laurette made a loud kissing sound, followed by an exhaled "haaaaaa"—the call of love, she explained, and they responded with more fortissimo musical wailing. The indri song, often compared to that of the humpback whale, is one of the loudest sounds made by an animal, one that peals through the forest for nearly two miles. It seemed the plaintive sound of a besieged species.

Laurette called the indri *babakoto* (loosely translated it means "ancestor"), and explained that they are *fady* to eat, as the people believe they are directly related to the animals. But even though it is *fady* to kill these in-laws and quite illegal, every year many within the reserve are lost to poachers, the "killer apes," who can traffic the meat for the comparative fortune of $1 a pound.

We finally sighted four of the boy-sized, tailless indris, two of each sex, with panda markings, teddy-bear ears, and haunting amber eyes. Their fur was thick and silky, predominantly black, with patches of white on their backs, rust on their tummies, and slated by gray around the bare, black muzzles. Four stories up, they ricocheted from precious hardwood to hardwood, through a palisade of trunks, one after another, as though catching the same bus to work. At one point the rain forest suddenly seemed to be weeping, and Laurette turned to me with a broad grin across her moon-shaped face. The indris were urinating, something they do in unison, and by schedule, every morning. They were obviously creatures of habit. Then they were off again, using powerful thrusts of long, agile, black-socked legs to make graceful, acrobatic flings. They looked so cuddly, so human, that it was easy to see why the locals believed indris were ancestors. It was difficult to fathom how they could be hunted for food.

I took the narrow-gauge train, nicknamed Fandrefiala after a long, slim forest snake, back to Antananarivo. A poster beneath the steeply peaked, dormered roof at Andasibe (Big Station) proclaimed FOREST, HEART OF MADAGASCAR, a propaganda piece designed to motivate a change in thousand-year-old habits. Yet on the way back the passengers' faces became soot-streaked from the trees burning out the windows, and flakes of forest fluttered through the car like pages of a yellowing book. Outside was a charred, ravaged wasteland, acres and acres of smoldering devastation. Much of the native forest was aflame from farmers engaged in the ecologically disastrous technique of slashing and burning new croplands, mostly to grow dry rice. Others were burning select eucalyptus and tamarind for *charbon*, charcoal. It takes about ten trees to create a 4-foot-high sack of charcoal which sells for about $1 a bag. "Forest in a Sack," Serge called it.

Left. Ringtailed lemurs strolling down the road in Berenty, the 1,100-acre private reserve near the southeastern tip of Madagascar.

While Madagascar is one of the richest nations in the world in terms of nature and biodiversity (the eighteenth-century French scientist Philibert Commerson called it "the naturalists' promised land"), it is one of the poorest economically, with an average annual per capita income of just $250. The country has been independent since 1960, yet under an old French colonial law still on the books contraceptives are illegal in Madagascar, so the typical rural family has eight or nine children. The country posts an alarming 3½ percent annual population growth—in the last twenty years the population has doubled to eleven million. In order to cook the rice necessary to feed so many children the parents need fuel, and the fuel of choice, because it is cheapest, is charcoal. Thus, for most in Madagascar, environmentalism is a long-sighted luxury that clashes with current crying needs; the immediate future, the next meal, is the priority. In 1986 over 40,000 people died in a famine in the southeast, and out the train window I saw countless children potbellied with malnutrition. But if alternatives to existing practices aren't adopted, the present will be a burnt sacrifice to the future. Incinerating one's own environment is ultimately self-immolation. Prince Philip, Duke of Edinburgh and president of World Wildlife Fund–International, recently watched a flaming forest in Madagascar and said to His hosts, "Your country is committing suicide." But even His Highness could offer no viable alternatives, no ready solutions. It is a tragedy without villains, a war against an enemy with no face.

I spent the night in Antananarivo, and out my window a blue haze of smoke draped itself around the flowering jacaranda trees, the spires of churches, and the roofs of the motley, orange-tiled homes. At breakfast I asked Serge if he'd had any luck in locating the restaurant. He hadn't, but he was still trying, and he hoped to find it by the following week.

In the warehouselike airport I lingered at a gift shop featuring belts ($50), wallets ($80), purses ($175), and briefcases ($300), all made from the skins of crocodiles whose forefathers swam from Africa during the Pleistocene. Now crocodiles have been hunted out almost everywhere on the island except in a few lakes in the north, where they are revered as ancestors, and sometimes as profitable components of luggage.

Opposite, top. On the road between Perinet and Antananarivo, the pastoral scene is interrupted by the smoke-plume of a man-made fire.
Bottom. "Forest in a sack." It takes about ten trees to create a 4′-high sack of charcoal that sells for about $1 a bag.
Below. A woman in Hellville, the main town of the "The Perfumed Isle," Nosy Be off the northern coast of Madagascar. The yellow mangary pigment paint is used to improve the skin, not as a decoration.

Opposite, top. Conservationists say 80% of the world's flowering plants (some 10,000 species) are found only on Madagascar and its satellite islands. *Bottom.* On the way to market in Fort Dauphin, the most beautifully located of all towns in Madagascar, bordered on three sides by beaches and backed by high green mountains.

***Oly*—a type of charm used as an object of worship.**

The next stop on my survey was the southeastern tip of Madagascar. Fort Dauphin, on the opposite end of the island, is 999 miles from Nosy Be. Fort Dauphin and San Francisco are geographical antipodes; I would be at the farthest point on the planet away from home. The flight at first took us over a puzzle of emerald-green rice paddy rectangles, then a faded landscape with a threadbare coverlet of green, the tattered felt of an old billiard table. But as we crossed the Tropic of Capricorn the felt faded to parchment, creased with chasms. The last minutes of the flight we sailed over a lacerated folding desert that poured into the sea at the country's oldest town, named after the Dauphin, the six-year-old prince who was crowned Louis XIV of France just as the first French settlement was being established in Madagascar in 1643. Once a thriving port, it is now a harbor in an advanced state of decay, and famous for its winds. True to reputation, it was dark and blustery upon arrival, with dark-bellied clouds pressing ozone into our nostrils. We quickly piled our luggage into a white Renault station wagon and headed north for the fifty-mile, two-hour drive to Berenty just as the last rays of the sun stippled the craggy granite mountains. Driving towards Berenty on a stormy night in October was like driving through a war zone. With the windows down we could hear the stutter of axes and the steady tears of saws between peals of thunder. On each side we saw fires lashing at the sooty night, burning to bring up new shoots of grass for goats and humped zebu cattle to graze. The lyre-horned zebu, more than a source of dairy and meat, are talismanic units of wealth, used to pay for marriages and funerals, and there is continued pressure to breed more, even more than the island's estimated ten million. These fires were singeing the edges of the wildlife reserve, perhaps burning away species yet to be discovered. When Indonesian explorers first paddled twin-hulled outrigger canoes to the island around A.D. 1000, Madagascar was cloaked in virgin rain forest. Since then a pygmy hippo, an aardvark, two species of giant land tortoise, three species of birds (including the *Aepyornis maximus,* or elephant bird, a flightless creature weighing 1,000 pounds and over ten feet tall that gave rise to the legendary giant roc of Sinbad's second voyage, according to Marco Polo, who heard the tale from Arab traders), and fourteen species of lemurs have disappeared, their habitat burned for farmland and fuel, overgrazed by livestock, and their doom hastened by overhunting. At least 238 of Madagascar's known species are now endangered, clinging to the sides of a shot and sinking ark.

Near "the tree where man was born," a bowl-bellied baobab, we took a right turn onto a pitted dirt track. It was a four-mile trundle through the species-unique spiny desert and across a neatly manicured sisal plantation to the 1,100 acres of private reserve owned by the wealthy de Heaulme family. These gardeners of Eden had decided to protect the various species of the area in 1936. A bony-faced, straw-hatted man wrapped in a blanket and wielding a spear was guarding the entrance, but as soon as he saw my white face he lifted the barrier and said, "Salaam." I was exhausted, so after dinner and a beer, I retired to a pillow filled with dry grass in my bungalow.

The next morning I awoke to a furry little dog-face with bright Bart Simpson eyes staring at me from the doorway. It was a ringtailed lemur. Then I heard the pitter-patter of little footsteps on the roof . . . dozens of them, more ringtails, scampering, tap-dancing, insisting I get out of bed. I pulled on my shorts and went outside to an alien panorama. There were scores of cat-sized prosimians with long snouts and squirrellike tails bouncing everywhere. There were three species within one hundred yards of my bungalow. Child-sized, creamy-furred, western sifakas were performing balletlike sideways leaps across the road; red-fronted lemurs were swinging like circus children from octopus

trees; and dove-grey ringtails, velvety tails curled like upright question marks, were frolicking around the grounds, snatching bananas from the accommodating *Homo sapien* visitors. Several lemurs had month-old babies clinging to their backs like miniature jockeys. It looked like a scene from the movie *Gremlins,* only here, on Main Street in Lemurville, all the critters were agreeable and nice.

Opposite, top. October is the burning season for Madagascar, when farmers take their torches to the remaining forests. *Bottom.* The lighthouse keeper on the tiny island of Nosy Tanikely.

At breakfast I met the Breakfast Club, a group of habituated lemurs who shamelessly seduce visitors into sharing the fare. I also met zoologist Dr. Alison Jolly, the doyenne of lemur-watchers, who was leading a group of twenty-five from Earthwatch (the Boston-based work-study group) on a two-week study tour. Alison began studying the lemurs of Berenty in 1963 for her Ph.D. at Yale. Now she is a professor of biology at Princeton. I asked Alison if I could join her for the day as she took her students around the reserve. She welcomed me, and took us down to one of her favorite haunts, the narrow gallery forest along the Mandrare River, an exuberant brushstroke of life on an otherwise bleak canvas.

It was a special day. With obvious delight, Alison explained that the females in the troops are dominant, and will win any battle of the sexes. "Just like home," one of the female students observed. Alison interpreted the animals' exhibited behaviors for us. Members of a troop of twelve snacked on the bean-like fruits of the tamarind trees, anthropomorphically sunned themselves, performed dazzling acrobatics, groomed one another, engaged in energetic play, marked territory boundaries with scent glands in the base of their tails, and sounded calls of alarm when a harrier hawk swooped overhead. I held a banana to one, who scampered over my shoulder from a bauhinia tree and peeled back the skin with human-like hands with black fingernails. As he chewed, his round, lustrous eyes stared at me like a little Rodney Dangerfield, begging for respect. Though the ringtail was only inches from my face I stared back across a vast evolutionary distance. Still, there seemed to be a flicker of recognition, and as I wondered how any caring human could kill and eat a big-eyed being who looked like a blood-related baby, the lemur seemed to look back and wonder as well. I reached over and touched the long, prehensile fingers of his little hand, and they were cold.

The *tandraka* is similar to a ground hog, but indigenous and peculiar to Madagascar.

Towards the end of the day, while lounging in front of a pit crawling with brown-backed radiated tortoises, I thanked Alison for a special day, and asked if she had heard of a restaurant that served lemur. She scowled. "If there is such a restaurant I would gladly go and burn the place down," she said, and stomped off towards her quarters.

The next day I was back in Antananarivo, where the morning mango rains had washed the streets but not the air. Serge told me he had found the restaurant and would take me there for lunch. It was Friday, when the world's largest open-air market takes over the streets, and so it was a slow drive to the Ambodifilao section of town, an area of decaying, pastel Gallic buildings with pointy roofs and second-story wooden balconies. The scene seemed a watercolor in the smoky, diffused light. There, on a narrow storybook street, was the Restaurant La Tulipe, subtitled Chez Claudine, with the tagline "Cuisine Chinosie, Specialites Gibiers"—*Gibiers* meaning "game." We were a bit late for lunch, arriving around 2:00 P.M., but there were a few stragglers, all European, finishing meals. We took a seat in the open-air section and looked at the menu. It featured *cuisses de Nymphe L'ail* (froglegs in garlic) and *Pigeonneau frits,* but no lemur, or "Tarzan," as the waiter called the animal. At first he claimed they didn't serve Tarzan. But then, after much insistence, he excused himself to speak with the owner. A few minutes later he returned to ask if we would like to try radiated tortoise, bats, or boas, none of which was on the menu. We said no, that we were really interested in lemur. He apologized and said they were out of lemur, but if we placed an order they could supply one by next week. That

Wooden staff decorating Madagascar tomb. Ancestors, or *razana*, have a very important place in the everyday life of Madagascar. There are elaborate rituals connected with the burial of the dead. Symbolic wooden sculptures like this decorate memorial sites. There is a custom called *famadihana*, or turning of the dead. Every few years bodies are exhumed, washed, and rewrapped in new cloth. The host family provides band and food.

wouldn't do, as I was scheduled to depart the country the following morning, so we shrugged our shoulders and decided to order *crabe farci* and zebu sandwiches.

Sometime into the first few bites the owner, Claudine, appeared at our side. She seemed quite sophisticated and friendly, a stout, jolly middle-aged lady with two gold rings on the fingers of her left hand and a gold broach in the shape of Madagascar on her breast. She looked vaguely European. In fact, she volunteered, her father was Chinese and her mother was half French, half Malagasy, and she owned the restaurant with her Chinese husband. Serge explained that I had travelled halfway around the world to taste lemur and that I was well connected in the United States; I could bring other connoisseurs who might enjoy the exotic tastes of her establishment if only I could sample the wares. She looked at me, searching for clues of sincerity, and suddenly broke into a radiant smile. "I think we may have some lemur frozen in the refrigerator, left over from last week. Would you like to try that?" Yes, we nodded. "Would you like it marinated in its own sauce, or with wine, ginger, and mushrooms?" Could we try both? Of course. She disappeared into the back.

A minute later I asked the waiter how we could tell it was really lemur she would be serving. He vanished into the kitchen, and returned with a fellow waiter. Together they proudly unfurled the skin of the red ruffed lemur, an endangered species. It had been killed by Claudine's Chinese brother-in-law, who was, they said, at that moment out hunting more. Minutes later the victuals were served, and a decision had to be made. I knew I wanted to track this restaurant down, I knew I wanted to uncover this small, perhaps symbolic atrocity and send word to the right people of these goings-on; but I hadn't thought of what to do if actually served the animal. I screwed up my face, then looked to Serge, who stared back with his one good eye. Then Claudine stepped to the side of the table and stared down at us. "*Bon Appetit.*" Then, after a beat, "What do you think?" I picked up the fork and picked at the tiny ribs. She hovered over us, and I took a bite. It tasted like tough beef, even smothered in sauce. I chewed, then tried the other plate, and Serge joined me. Claudine stood over us for several chews, and we smiled thinly throughout, then she finally excused herself, saying she hoped I would spread the word of her haute cuisine and send her more tourists. I nodded a promise, and continued to chew, and when she left the room I took my camera from my pocket and took photos. When it came time to pay the bill I couldn't help but notice that Tarzan was the heftiest item, more than double any other dish. It was 6,000 Francs Malagache, about $4 in U.S. currency.

I can't say if it was the lemur or the crab, but I felt ill the rest of the day, especially as we toured Parc Tsimbazaza, the national zoo, and the keeper showed me the red ruffed lemur. He explained that there were presently more in captivity than in the wild as they were being poached so effectively. But by evening, after a frosty glass of Three Horses beer, I took pen to paper, and soon felt better. I wrote a note to Alison Jolly, who would be in Berenty for another couple weeks, and in it I told her of the restaurant at 17 Rue Rabezavana. I told her if she were going to burn the place down, I would gladly supply the match. It was the one place in Madagascar that deserved to go up in smoke.

Sri Lanka

The Fool on the Hill

On leaving the island of Andoman and sailing a thousand miles, a little south of west, the traveller reaches Ceylon, which is undoubtedly the finest island of its size in the world. . . . In this island there is a very high mountain, so rocky and precipitous that the ascent to the top is impracticable, as it is said, except by the assistance of iron chains employed for that purpose. By means of these some persons attained the summit, where the tomb of Adam, our first parent, is supposed to be found.
—Marco Polo, 1293

Like the cheek of a young girl, the sky began to blush. Faint rows of primrose light soon became golden bars, through which the dawn glided out across the jungle horizon. The stars grew dim and dimmer still, until at last they vanished; the citron moon paled, and its mountain ridges stood out clear against its regal face. Then came spear upon spear of light flashing far away across the boundless wilderness, piercing and firing the veils of mist until the rain forest was draped in a tremulous glow.

As the last drops of darkness drained from the sky, a supernatural tableau presented itself. The rising sun behind me threw a Himalayan shadow of perfect triangular proportions across the mist-spun awakening countryside. The umbra, a huge, dark cone, seemed to fly across the vista, cast by the mountain upon whose peak I stood, a mountain that towered over the pear-shaped island like a pyramid hewn by distant gods, very large gods. It was the magnified morning shadow of this rock and jungle promontory that superimposed itself for miles over tea plantations, villages, and folding ridges, creating one of the most awesome natural sights on earth. But wait, it got better: The "Specter of the Brocken," a rare event indeed, raised its curtain—my own immensely amplified shadow stretched over the vaporous air below like that of some Brobdingnagian visitor, and it floated on far-off wraiths of mist, looped by a rainbow halo. As quickly as it all appeared, the shadows of peak and person began to shrink. The blue adumbration galloped towards the base of the mountain as the sun rose behind me, lightening the sky from lapis to sapphire. In seconds the shadow had vanished into the bedrock, and butterscotch sunlight poured over the panorama all the way to Colombo and the plate-blue sea.

I opened my eyes and was slapped by the thick, cold fog that wrapped around me like a wet sarong. Visibility: about four feet. For all I could tell, I was swimming underwater in a colossal cup of chilled cream soup. The most distinctive sight was the fingers of my extended hand, and then just barely.

It was precisely for this special shadow that I had climbed Adam's Peak, the most famous piece of landscape on Sri Lanka, the tiny island, formerly Ceylon, that hangs like a teardrop off the southern cheek of India.

The only problem was I didn't see it. Nothing described in the first two paragraphs happened in my sight. In my imagination, yes; I saw it all, in Sony Trinitron color,

Opposite. The blue adumbration known as "The Specter of the Brocken," as seen at sunrise from the summit of Adam's Peak.

A Yaksha. One of many demons that occupied ancient Ceylon.

drawn from the descriptions of pilgrims and guides, and I saw it many times as I took step after grueling step upwards in the India-ink darkness. The timing was perfect. I began the trek at 1:00 A.M., as hundreds of thousands of the devout have done before me, and I arrived at the sacred summit sixty minutes before sunrise. But there was nothing to see—no shadow, no vista, no phenomenon—just silence and soup. And the footprint.

This is not the tale of a first ascent, nor a technical climb, nor even an assault to a country's highest point. Adam's Peak is just 7,360 feet (2,243 meters) high, and is Sri Lanka's fourth-highest mountain. (Pidurutalagala holds the crown at 8,281 feet/2,524 meters.) Yet this molehill of a mountain has seen more climbers than any pitch in the Sierras or the Andes; and more have died on its slopes than Everest, McKinley, Kilimanjaro, Whitney, K-2, the Eiger, Nanga Parbat, and a dozen other classics combined.

We trundle into the village of Dalhouse, in the heart of Sri Lanka's tea country, just past midnight, unaware of how grossly ill-prepared we are for the ordeal ahead. How hard could the 4½-mile trek up this mountain be, a journey made by thousands of pilgrims each year, many of them old and in poor health? That was the collective cognitive ignorance as we set out for our little expedition. During the pilgrimage season, the dry months of December through March, the path is illuminated by a string of electrical lights, and from a distance it is said the sight of the endless progression of pilgrims, many with their own torches, appears as a river of light flowing upstream, something like a backwards lava flow. Rest stations, called *ambalamas,* dot the trail and offer shelter, tea, fresh fruit, and water. On a full-moon holiday there are as many as 20,000 pilgrims slowly stepping upwards, and it can take a healthy hoofer eleven hours or more to reach the summit.

But now it is late May, the southwest monsoon season, and the mountain is dark

Below. A Buddhist monk meditating in the Cave of Rawana at Ella Rock.

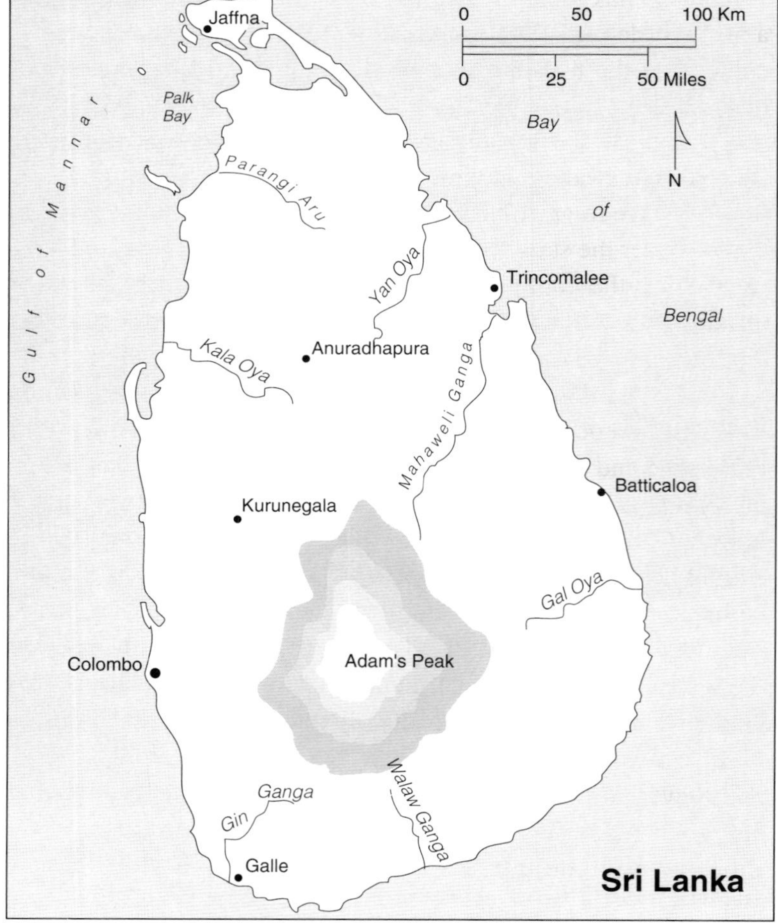

and empty. The only light in any direction shines from a hut near the start of the path. This doesn't faze our intrepid group, however. We file off the bus into a tepid drizzle with the insouciance of Sunday picnickers. Tilak, our usually stodgy Sinhalese (Buddhist immigrant from North India) guide, wears a shower cap. He looks silly, but who can talk? Nadya, from the Ceylon Tourist Board, has wooden clogs, no socks, two T-shirts, and nightblindness. Branka, a Yugoslav travel agent, wears a chic silk blouse, skirt, sandals, and two towels pinched from the hotel. Pam, our photographer, has a cold and hasn't slept in two days. Leah, fifty, mother of four, is the best dressed in windbreaker and tennis shoes. Leo, thirty-one, from Oklahoma, at the beginning of metabolic decline with a hint of Buddha belly, at first shunned the opportunity to join this expedition. It was a foregone conclusion that when we boarded the bus in the ancient highlands capital of Kandy, Leo would be back in bed sound asleep, the sane way to pass a monsoon night. Sanity notwithstanding, there he was on the bus . . . grinning. Leo, like me, has running shoes, a pullover, and long pants. No hat. No Goretex. No wool. Amongst us we have no raingear, and two flashlights.

Tilak and I cross over to the hut where a bare lightbulb swings in the wind, and a silhouette of a man appears in the doorway. The stranger speaks in Sinhala, the primary language of Sri Lanka:

"You can't go up the mountain now!"

"Why?" Tilak asks.

"It is dangerous. It is monsoon. But worse, as there have been no people, the wild animals have taken over the path."

"What animals?"

"The leopards. Wild boar. Cobras. Only fools would try the mountain now."

Tilak looks worried. And worried he should be. Leopards have killed more than a few people in the highlands. Wild boar can claim a few lives as well. And the Ceylon cobra is of the five-step variety—one bite and you can make five steps before dropping dead. Just the day before, one crossed Pam's path on a day hike.

The Dance of the Gale Deviya, God of the Rock. In ancient Ceylon there were yearly celebrations, with dances on the hills. There are specially named rocks where this deity danced.

Still, Tilak is here to guide us, and we want to hike the peak, to see the famous shadow. He swallows, wipes his glasses, turns to the wild bunch and says, "Let's go."

We're at 4,000 feet and still climbing. Like a cast-iron griddle over coals, the clouds cover the stars. Our two flashlights burn holes in the darkness, and we stumble and stalk along the cobbled path. We pass a fountain used for bathing, where the pilgrims cleanse themselves before supplicating to their inspiration at the summit. We decline the dip, content with the shower, and continue to trip upwards.

The rain intensifies, stinging exposed skin. Beneath us stone gives way to wood as we cross a decaying planked bridge that muffles the thunder of cascading water. I stop midbridge and scan the churning stream with my light. It ignores my intrusion, as it does all who pass, and spits at the lowlands ahead. This young stream is the start of the Mahaweli Ganga, Sri Lanka's longest fresh-water course. Springing from the flanks of Adam's Peak, it is 207 miles from source to sea. It's wild and restless here, as are its many siblings. But far downstream, near Kandy, the Mahaweli is bound and gagged, broken by the massive mile-long, double-curvature Victoria Dam—the fifth dam to be flung across the water, and a project that hopes to bring the tiny island closer to self-sufficiency. On another nearby tributary, the Kelani River, the classic film *The Bridge Over the River Kwai* was shot; on other feeders to the great Mahaweli Bo Derek's *Tarzan the Ape Man* and Steven Spielberg's *Indiana Jones and the Temple of Doom* and Duran Duran's video Hungry Like the Wolf all found the exotic backdrops Hollywood could never provide. A drop hits me in the eye, and I turn and continue.

Above. One of the fishing fleet at sea off Negombo on the West Coast.
Right. Nestled in the foothills of the Hill Country on the banks of a lovely tree-lined lake is the city of Kandy, the center of traditional Sri Lankan culture.

Shortly we come to a huge arch straddling the way. Nearby sit two shrines, one with a porcelain Buddha, the other with the elephant-headed Hindu god, Ganesha, the deity who decides between success and failure, who removes obstacles or creates them as necessary. These shrines are the first evidence of the polytheistic paradox of The Peak. The mountain is singular in its holy significance to four major faiths. There is nothing quite like it anywhere, and millions have trekked countless miles to pay homage to the footprint shrine atop the mountain.

The print itself is said to have been discovered by the first-century King Valagam Bahu while he was in exile after being driven out by the invading Tamils. He was wandering through the forest when he met a doe-eyed deer who lured him to the summit, then vanished at the footprint site. Valagam knew he was onto something, and once back in power ordered a shrine erected over the print.

Marco Polo, from the title page of the first printed edition of *Travels of Marco Polo*, 1477. His travels opened the civilizations of the East to the Western world.

Opposite, top. Tea pickers at the Mousakellie Reservoir near the base of Adam's Peak.
Bottom left. Even the palm trees have sensuous attitudes on the sultry coast of Sri Lanka.
Bottom right. Stilt fisherman off Weligama. They sit on crossbars tied athwart forked branches of trees planted in the sea bed.

Sri Lanka, with the Sinhalese constituting 74% of the population, is one of the half-dozen predominantly Buddhist nations in the world. *Sri Pada,* which means "The Sacred Footprint," is the Buddhist name for the mountain; followers believe Siddhartha Gautama Buddha himself planted his foot on top some 2,000 years back.

Hindus call it *Shivan Adipatham* ("the creative dance of Shiva"), and believe the lofty shrine preserves the pedal impression of Lord Shiva (the supreme Hindu god, and brother to Ganesha). Muslims, who inexplicably found the name that sticks, claim it was Adam who left the mark when he was expelled from Paradise and forced to stand on one foot for a thousand years. Finally, the Christian Burghers, descendants of Portuguese, Dutch, and English colonialists, posit St. Thomas, the early apostle who preached in South India, as the man who stood on the summit weeping bitterly for his doubts with heavy heart and foot. Then there are those who theorize it was the extraterrestrial equivalent of a Sierra Clubber, visiting our earthly wilderness and leaving only a footprint.

Whoever or whatever is to be believed, the peak is, in the words of British author John Stills, "one of the vastest and most reverenced cathedrals of the human race." And here we are irreverently slogging up its face.

Slowly, almost imperceptively, the grade steepens, and my flashlight begins to fade. Then the sky lets loose—rain slashes us and my trousers feel as though they've been weaved with wet cement. Pam spots a light ahead, and leads the charge to shelter—an overhang at the entrance to the Japan-Sri Lanka Friendship Dagoba. We've come one mile—the easiest—up the trail.

"This is ridiculous," someone cries over the storm. "Let's turn back." No one argues.

We stand shivering, surveying the situation and each other. Tilak breaks out a bag of ham-and-cheese sandwiches, and I devour one before I remember the taboos cited by a Buddhist monk back in Kandy:

"If you climb Sri Pada you must not inquire as to the time or distance to the top; nor can you consume alcohol; nor eat meat."

I look to where I imagine the summit soars. "Sorry, Siddhartha."

The group wants to repair back to the bus; that seems perfectly logical. But I see a couple of stars wink through the mantle, and I feel drawn to them.

"I think I'll make a go for it," I announce softly.

"I'll join you," volunteers Leo, who originally had the gravest misgivings. I borrow some batteries from Pam's camera flash, switch them to my flashlight, and we part.

The rain lessens as the ascent steepens. I sweep my light along the trail and see a set of crude buildings. One is marked in handpainted letters: PUBLIC HEALTH OFFICE AND CORONER. A seemingly odd place for a government office, until the numbers and health of

Below. The Elephant Bathing Place at Katugastota near Kandy. At the foot of the main bridge mahouts lead their great beasts to the river to cool off after a day's labor. *Opposite, top.* Husking coconut fiber. The plaited fronds become roofing material for village huts, and the fruit's fiber goes into the making of coir rope. *Bottom.* The view from the Trident Hotel, overlooking the Indian Ocean.

the climbers is considered. A Victorian-age guidebook to Ceylon describes the hike:

> Others struggle upwards unaided, until, fainting by the way, they are considerately carried with all haste in their swooning condition to the summit and forced into an attitude of worship at the shrine to secure the full benefits of their pilgrimage before death should supervene; others never reach the top at all, but perish from cold and fatigue; and there have been many instances of pilgrims losing their lives by being blown over precipices or falling from giddiness introduced by a thoughtless retrospect when surmounting especially dangerous cliffs.

With a new note of caution, we continue our flight. The clouds cover the last patch of blue, and wring out more rain. Every thirty steps or so Leo stops to rest, and I with him. He's winded, very much out of shape, but I appreciate the breather and am content with his pace. At one rest I turn back downwards and see the faint outline of the Dagoba, penciled by poor floodlights. It looks like a pale spaceship, which reminds me. . . .

Ganesha, the Hindu elephant-headed god of wisdom and good fortune.

Arthur C. Clarke has been Sri Lanka's most celebrated resident for over twenty years. The author of a raft of sci-fi books such as *2001: A Space Odyssey,* he took this trek some time back and was sufficiently impressed that he wrote a book centered around Adam's Peak—*The Fountains of Paradise.* The story oscillates between the first century, when the pilgrimage is pure, and the twenty-first, when the faithful, who still seek merit by climbing unaided, compete with a tourist tram and the plans of a master engineer to build a space elevator atop the peak.

At present, not even a helicopter carrying the prime minister can gain access to the peak. But in Clarke's story science/progress prevails—for a price. If active plans are successful, a feature film of the story will soon be shot here. Echoing Clarke's story, some of the faithful have already vowed to prevent it.

We continue to climb.

Somewhere in this middle distance we're joined by Mr. D. He trots beside us, then bounds a few steps ahead, then waits. Whenever I switch off my light at a rest stop to preserve the batteries, he whines. We discover Mr. D's not really a he when he squats for

relief, but Leo's name for the medium-sized mutt sticks, and she becomes a full-fledged member of our expedition.

Mr. D is good companionship. She is always urging us upwards, and seems ever pleased, as long as my light shines. And she speaks about as much as Leo, which is fine with me. This somehow doesn't seem the place for snappy conversation; thoughts should be directed inwards. I think about Peter Matthiessen's book, *The Snow Leopard,* in which he treks tirelessly through the Himalayas for a rare cat he never finds, all the while stitching Buddhist revelations through his ruminations. I search for something profound, of heretofore unplumbed significance, as I move my legs in existential exercise: the inner game of hiking, Zen and the art of sole maintenance. But this is a confusing mountain. There are some bizarre metaphysics afoot. Should I seek Buddhist enlightenment, which speaks of suffering and discipline, certainly applicable here? Or Hindu Brahminism, which demands a matter-of-fact repetition of gestures, again relevant? Or Islamic Muhammadanism, which promises a higher place if a *hajj* is undertaken? Is this a stairway to heaven, or a Tower of Babel? My mind blanks. I lift one leg, then the other. It's beginning to get cold, and I find it difficult to believe we're just five degrees north of the Equator.

"How far do you think it is?" I blurt, realizing before Leo can answer that I've committed my second sin, asking the distance. Leo grunts a non-reply, and my flashlight fades to black. "No wonder Firstlite went bankrupt," I curse. It's dead black. Mr. D starts to whine, and I stumble off the path into the slippery jungle. "What's that?" Leo's voice cracks at a rustling of leaves near me, and we both think leopard. Only silence screams back.

I have one last set of batteries, those in my camera flash attachment. I sit down and circumspectly begin the operation, using touch and memory to keep the positive and negative leads straight. It's such a simple task, but my fingers are numb and obvious parts seem unfamiliar. Even though we're sheathed in almost complete darkness, I squeeze my eyes shut to concentrate on the task. It's a silly chore, but I think of others that have taken on enormous importance in the unanticipated setting—trying to ignite a fire with wet matches on an Arctic night; trying to fashion an oarlock from sticks after a disastrous capsize on a wild river. A piece of cake by comparison . . . a small mountain, just a hill really, and inclement weather, but nothing life-threatening. Mr. D lets out a howl, and the light spits on.

Renewed, we continue the march. It's steeper now, with more switchbacks. I notice the litter. No poptops, no Mountain House or Trail Mix wrappers here. Instead, incense containers, pineapple rinds, orchid petals . . . and white thread, miles of it, strewn in filigree patterns along the sides of the path and in the trees. We're at Indikatupana ("place of the needle"), a sacred stopping point that demands the devout to fling a threaded needle into a shrub adjacent to the trail, marking the spot where Buddha is said

Entering the forest of the god,
Beholding the god's auspicious forest,
Bestowing merit on the god with cheerful heart,
Ever making him our refuge;

Buddha first all men adore,
They travel many a yodun in the forest,
Thinking of Saman Deviyo as they come,
'By virtue of these merits (that we give) the Buddhahood thou shalt attain.'

Go we all to worship Samanala,
And overcome the ocean of rebirth,
Let us offer these flowers and lamps,
Let us sing all through the forest!

—Verses from the *Sri Paqa Mana*, a popular song about climbing Adam's Peak

Opposite, top. The Bandulla Falls in the high country. *Bottom.* Budding Buddhists. *Left.* Residents of the government-supported Elephant Orphanage.

to have paused to mend a tear in his robe. It's an excuse to rest, and we do—but we ignore the rite of passage.

Upwards again, with Mr. D leading the way. A saffron butterfly flutters through my flashlight beam, then drops dead at my feet. Legend says the butterflies are the lost souls of soldiers who died centuries past in battles near the mountain's base, and that they, too, make a final pilgrimage here. In fact, an earlier name for the peak is Samanala Kanda, "Saman's Hill," after the deity Saman, who was regarded as supreme before the advent of Buddhism. Now, when clouds of yellow butterflies tremble around the mountain each year, it is called *Samanalayo.*

With exaggerated slowness we trek. I glance over at Leo. He's bent over like an old man, his sides heaving. He staggers another dozen steps before collapsing in exhaustion. I think of films I've seen of climbers on the final pitch of a major peak, the measured, painfully retarded steps, the languid sway of the arms, the wasted eyes, and I see Leo. Nonetheless, this is just a hummock compared to its Himalayan sisters, just a fourth as high as the great ones.

The grade is so steep it's like climbing a ladder now. I see the number "50" on a step, and shoot a glance at Leo. "I wonder . . . ."

Fogged, incoherent eyes peer back. I count the steps out loud, "49, 48, 47 . . . ." Leo joins in. "33, 32 . . . ." The path zigs to the right. "27, 26, 25 . . . ." Then to the left. "21, 20 . . . ." The trees part, and at 5:05 A.M. we confront the emptiness of the summit.

Off to the side, a few steps from the tiny crest, a bulb burns through the cracks of a ramshackle hut. A door creaks open, and a middle-aged man rubbing his eyes emerges.

"*Karunauai,*" he says, the traditional greeting passed between pilgrims. It means simply, "Peace."

He gestures us in for tea, and we accept with alacrity, desperately cold and tired and knowing that sunrise is still an hour away. His home is a six-foot by six-foot shack, bare save for some blankets, a space heater plugged into a compact gasoline generator, a frayed-corded jambox, some tins of dry goods, and a wooden bunk with no mattress. Though our host doesn't speak English, he communicates that his name is Yureratna Sri Balaabdullah Ratnapura. He spends six months a year atop the mountain as a lonely sen-

There is a universal Hindu love of trees and forests. Sometimes tree gods are malignant, sometimes loving. Trees are often drawn with a face among the branches to show the tree spirit. Top: cannibal king seated beneath the tree, with a thorn in his foot. Above: human offering to the tree *devi* (or god).

Opposite, top. Fishing boats near Kuchchaveli.
Bottom left. During the Buddha's lifetime he is said to have transported himself to Sri Lanka on at least three separate occasions.
Bottom right. A temple monkey picks lice from its newborn baby's coarse hair.
Left. The Sinhalese today believe that they are a chosen people destined to protect and preserve the Buddhist faith in their island home.

try; the rest of the time at the place of his birth, the nearby "City of Gems," Ratnapura, his surnamesake.

As Yure serves up the hot, sweet tea, the clouds outside discharge the angriest torrent yet. It sounds like Armageddon outside; inside it feels as though we're in the snare drum of a heavy metal band. Leo digs into his pack and pulls out some soggy cheese sandwiches, which we share with Yure. We toss one out the door to our leader, Mr. D. We still have thirty minutes to showtime, so we warm our feet by the heater and listen to Buddhist prayers over Yure's radio.

"Did you cut yourself?" Leo points to my foot. "What do you mean?" I look down and see a sock soaked in fresh blood, but I don't feel a thing. Rolling up my pants leg reveals a rivulet of shiny blood coating my skin, but no cuts or scabs. Farther up my thigh I find the culprit: a liver-brown bloated leech, probably hitched on since I stumbled into the side of the overgrown path. Leo checks too and also finds one of the little suckers. Then he checks his watch: 6:05.

"It's time."

Makar represents Capricorn in the Hindu zodiac, and is also used on the banner of Kama, the Indian god of love.

We pull the parasites off and move to the door. Since shoes cannot be worn at the summit temple, we merely hook our camera bags over shoulders and stumble outside. Though the storm has subsided, it is bitterly cold. My bare feet stick to the surface as though to ice, and my face is buffeted by the brutal wind. The night dissipates to dawn as we take the final ten steps to the temple, Mr. D bounding ahead. And there is fog, thick and impenetrable—as we make the final step to the zenith of our journey it is quite clear we'll see nothing of the famous vista and its shadow. All we can see is Mr. D wagging his puny tail. Leo shakes his head. "I can't believe I just climbed 5,000 steps just to see a dog's bare ass."

But there is more. Four rusted, cast-iron bells mark the four corners of the shrine, and Leo and I each ring one once, as is tradition—a chime for each pilgrimage. Then we step to the small chamber at the apex. It is elliptic in form, surrounded by a five-foot-high parapet. An iron door rolls back to reveal the *Maluwa,* an outhouse-sized room that preserves the Sacred Footprint under a size-40 indentation. Rupees scatter the hollow, dear offerings to whomever made the lasting impression. Leo digs in his pocket, then tosses a quarter alongside pieces that equal 4¢ U.S. each. Certainly, though, the quarter's relative worth as an offering, coming from a relatively affluent Westerner, pales to the rupee in a country where a laborer commonly makes 40¢ a day.

We turn and face the fog. I wonder if committing the sacrileges on the ascent turned the view against us.

Simha, the lion. The lion is a mythical ancestor of the Sinhalese, standing for power and majesty.

In a way, though, I feel glad we are denied the sensational sunrise, that there is no spectacular shadow show. There is appeal in the ordeal, and that is what we had undertaken—no prize for participation; no passes to heaven or Nirvana. Only inner merit for achieving a personally fabricated goal, for journeying from A to B. Without the view our effort is little more than exercise, but it feels good. Even Mr. D lets out a satisfied yelp that spills into the mist.

Days later, back at a sea-level cocktail party, I meet the venerable H. P. Siriwardhana, Chairman of the Ceylon Tourist Board, a lofty position in a country whose fourth greatest source of revenue is tourism. I ask him if he has ever climbed Adam's Peak, and he blanches, as though I'd asked a Colombo monk if he'd ever adhered to the Dharma.

"My son," he replies after a long pause. "There is a saying in Sri Lanka: Only a fool lives on this island and does not climb Adam's Peak at least once . . . and only a grand fool climbs it twice."

I knew then I wanted to be a fool of grandeur. Besides, I still hadn't seen the shadow.

Lombok

The Place No One Knows

"Pass in, pass in," the angels say,
"In to the upper doors,
Nor count compartments of the floors,
But mount to paradise
By the stairway of surprise."
—Ralph Waldo Emerson, 1847

It was sunset; not a soul in sight. Tiny pieces of coral tinkled like chimes in the lapping waves, cool waves that softly rolled and licked my toes. The wind played in the palms, sighing like a deck of cards being shuffled. Across the waters, a volcano seemed to simultaneously lift the horizon and pull the red sun down. The air was as thick and sweet as the juice from a passionfruit. This fantasy was real, only I was in a place no one knew. If there is a Valhalla, at that moment I felt I had found it.

This nonfantasy island is a place where visitors arrive by pigeon, where "air" is water, where water is served in a plastic bottle with a label promising better health, where the largest tour company is called Swastika Travel. A cryptic religious cult survives on a volcano and worships holy eels, and Western traditions and rules are coiled and plaited, spun on a different axis, so life seems one continuous phantasm. It is a place with a singularly unromantic name, sounding more like a lower back muscle, or a New Age running shoe, than the beach paradise promised in brochures. Such brochures are hard to find, and few travel agents, tour operators, or even geographers could find this mote on the map, though it swims in the South Pacific in the shadow of one of the world's most famous paradise islands. When it is talked about among the cognoscenti, and the occasions are few, it is inevitably compared to The Famous Island to the west, just fifty kilometers across the strait—Bali. The Famous Island is like an excellent song played too often on the radio. It has been praised, hummed, dissected, touristed, filmed, and written about with more words than grains of sand on Bali's Sanur Beach. Its uncelebrated neighbor, though it has longer, whiter, and wider beaches, a higher volcano, and clearer waters, suffers from anonymity, and envy. It has never been featured on *Lifestyles of the Rich and Famous*.

I first heard of Lombok while watching director Peter Weir's 1982 film, *The Year of Living Dangerously,* a romance tale set against the incendiary summer months of 1965 in Indonesia, when communists were plotting a coup, and President Sukarano was about to take his fall. Somewhere in the middle of the film Mel Gibson, playing an Australian foreign correspondent stationed in Jakarta, is praised for a piece he did on Lombok. "I found it a bit melodramatic," co-star Sigourney Weaver taunts, and sets the stage for a steamy passion play between them.

Opposite. Moonlight over West Lombok, the magical and little-known Indonesian island on the edge of the Wallace Line.

Below. A Lombok boy gazes across the Lombok Strait to the richer and more famous Bali, just thirty miles away.

Where is this Lombok, I wondered, this island whose decidedly unamorous name helped ignite an epic tropical affair? In a kind of "Quest for Corvo," I had to find out.

Six years later I found myself eight degrees south of the equator and one hundred miles east of Java, puttering over the shimmering waters of the Lombok Strait, the deepest stretch of water in the Indonesian archipelago, in something called a CASA 212. It was a white 24-seat twin-engine flying shoebox, a plane built jointly by Spain and Indonesia, and owned by Merpati Airlines, a subsidiary of the government-owned Garuda Indonesian Airlines. In Indonesia, Garuda is a legendary eagle, and the proud bird is the international carrier serving Paris, London, Tokyo, and Los Angeles. Merpati means pigeon in Indonesian, and the puddle-hopper serves remote villages in the middle of Borneo and New Guinea. It also serves Lombok.

The 25-minute flight was $9. That included pieces of hard candy, to help ears pop in the unpressurized cabin, and a stunning approach view of the razor-backed, violently green, mist-wreathed island.

As we touched down at Selaparang Airport (named after a fourteenth-century Indonesian kingdom), we taxied down a runway whose grassy shoulders were being trimmed by bent-over women in big woven toadstool hats wielding tiny hand scythes, similar to the ones use in rice paddies throughout Asia. This was the first sign things were changing in Lombok. Ignored for centuries by those beyond its shores, its residents lived and worked around rice planting and harvesting, rarely thinking about the needs and economies of visitors. Now, in an effort to evoke a positive first impression for the tourists now discovering the island, the government was pulling people from the rice fields and teaching them to groom and spruce.

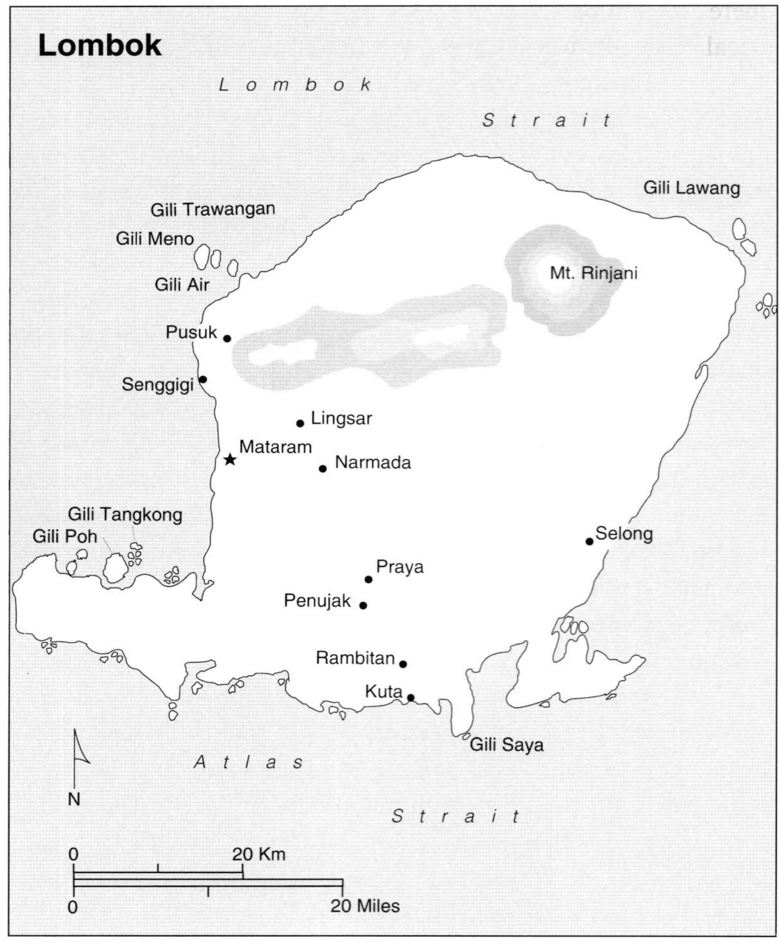

There are seven tour operators on Lombok, up from zero five years ago. I couldn't bring myself to hire one called Swastika, even if it was the largest, even though the Swastika, in local Hindu lore, is the ancient symbol of the creation of and continued life in the universe. So I hooked up with Bing-Bidy, a successful forty-eight-year-old Chinese cabinetmaker. Two years ago, Bing-Bidy sensed the coming opportunities in tourism, and went out and bought three used Toyota vans, recruited six waiters who spoke English to be his guides, and printed up cheap fliers announcing Bidy Tours and Travel. Bing-Bidy, who doesn't speak English, met me at the airport with his chief guide, Surya Kukuh, and Nanang, his driver. After much hand pumping and exchanging of pleasantries, we piled in and trundled up the waringin-tree-lined, narrow paved road some ten kilometers north to the Garuda-owned Senggigi Beach Hotel, the leading property on the island.

The Senggigi Beach Hotel is reason enough for making the trek to Lombok. Opened in April 1987 with fifty-two rooms, it is situated on perhaps the most scenic beach in the world. Backdropped by a rim of tropically dressed mountains, camouflaged by groves of coconut palms, it straddles a peninsula of sugar-white sand and glass-clear water. Due west, just across the way, is the overwhelming majesty of the great volcano, Gunung Agung, 3,140 meters high, and the mother mountain of The Famous Island. Every evening, on cue, the sun sets behind this Hollywood matte.

An ancient bas-relief carved out of a block of lava. The Hindu goddess Durga, "the Exalted Virgin," is portrayed on the back of a kneeling bull. She has eight arms. Her lower right hand holds the tail of the bull, while the corresponding left hand grasps the hair of a captive, Dewth Mahikusor, the personification of vice, who has attempted to slay her bull. He crouches at her feet in supplication. The other hands of the goddess hold, on her right side, a double hook or small anchor, a broad straight sword, and a noose of thick cord; on her left, a girdle or armlet of large beads or shells, an unstrung bow, and a standard or war flag.

Just past the open-air lobby is a clover-shaped pool, with a sunken swim bar roofed by a replica of a traditional rice barn. Beyond the pool is a deer park and aviary, which in turn is just a few steps from the beach. The bungalows are early Club Med crossbred with Holiday Inn. Fragrant frangipani trees hang over the back porches. There are no phones in the rooms (the management will arrange for a wake-up knock on request), yet there are minibars, air conditioners, and televisions featuring two videos a night. Electrical blackouts were routine, and somehow added to the charm of the place—a comfortable, unbelievably scenic retreat distinctly out of pace, the perfect rendezvous site for a Mel-Sigourney tryst.

Yet there was a sense that this magic, this primitive cool, was precariously balanced on the edge of a precipice. Lombok is still relatively undiscovered. The beaches are empty; it's hard to find cheap souvenirs. There are no discos or massage parlors or fashion emporiums. You can't parasail or jet-ski. You can't even buy a decent hamburger. But it's a window in time that is about to be shut.

Putu Swandi, the Senggigi's manager, boasted about how clean Lombok's beaches were compared to The Famous Island's, where tourists and regular festivals leave litter in prodigious amounts. Yet in the next moment he proudly pointed to a local woman carrying a yard-high pile of bricks on her head, a member of a massive construction team working on the hotel grounds. Another eighty rooms were due for completion by June; 150 more by October. The hotel had close to 100% occupancy since opening just a year before. Next door, a competing 200-room property was in the final stages of assembly, and more ambitious hotels and restaurants were planned around the island. Lombok's chief attraction was that it was not like over-touristed Bali, but there was a pell-mell movement to attract ever-larger numbers of visitors. Something would have to give.

Left. The revered Segara Anak, the holy crater lake near the summit of the Rinjani volcano.
Top. The sleek one-man *jukung* is small enough to navigate the tight coral passages of Lombok.
Above. Women in Lombok traditionally assume much of the heavy labor, including carrying charcoal to market.

The following morning my thirty-five-year-old barefoot guide, Surya, took me on a tour of the southern side of Lombok. Surya looked different from Indonesians on Java or Sumatra. He was darker, his hair was straight, cheekbones high, and his neck short. He explained he was *Sasak,* the original and predominate tribe on the island, a tribe who numbered almost two million out of a population of about 2.5 million. *Sasak* is the name of a type of bamboo raft once used to cross the Lombok Strait, and legend has it Surya's ancestors arrived on Lombok by these rafts from Burma or northwest India centuries ago, while the lighter-toned people populating the Indonesian islands to the west migrated down the Malay Peninsula. Surya went on to explain that while today virtually all Sasaks are Muslim, originally they were animists who believed in the innate liveliness of inanimate objects. Sometime in the early 1600s, though, the island was attacked from two fronts. Muslim traders from the Spice Islands established colonies in the east and converted the Sasak aristocracy to Islam, while Hindus from across the Lombok Strait made claims on a western section of the island, a section once used as a penal colony by the sovereigns of The Famous Island. Ruling power passed back and forth between the two religions for almost three centuries, until a Dutch invasion in 1894 officially turned Lombok into a Christian enclave, but not without a bloody fight. When it became certain the superior Dutch forces were going to overpower Lombok, hundreds of resistance fighters, including members of the aristocracy and royal family, committed a ritual mass suicide attack called *Puputan,* in which they deliberately marched into the lethal fire of Dutch artillery.

An early Sumatran youth.

The Dutch were finally ousted with Indonesia's independence in 1945, and today just 1% of Lombok is Christian. Another 15% is Hindu, and the rest is pretty much Muslim, except for an uncertain percentage, Surya guessed about 3%, that practices an odd and unique conglomeration cult called "Wektu Telu." It is almost a secret sect, unofficially recognized by the government, and practiced by unknown numbers of peasants in the remote mountains of the north. "Wektu Telu" literally means "three results," and is the result of combining the most attractive elements of three religions: Islam, Hinduism, and animism. The fundamental tenet of the faith is that all important aspects of life are underpinned by a trinity: the sun, moon, and stars; Allah, Mohammed, and Adam; the three rice crops their land can produce in a good year. Unlike orthodox Muslims, who pray five times a day, the Wektu Telu bow in prayer on just three occasions: every Friday, the Muslim holy day; on Idul Fitri, the festive day that ends the solemn fasting month of Ramadan; and on the Prophet Mohammed's birthday. They don't believe in mosques, or any man-made edifices specifically built for worship, and they don't practice the *hajj,* the annual trek to Mecca undertaken by Muslims with deep faith and pockets. They believe in meditation and their own spirits, some Hindu in origin, others pagan, and they eat pork with impunity.

I asked Surya how I could meet the Wektu Telu, and he explained that their few remaining villages were difficult to reach. Even if we did stumble into one, true Wektu Telu would deny their faith, as government prejudicial policies had turned the religion into an underground congregation. It had formally disappeared in 1968, though everyone knew followers survived in pockets in the high backcountry. The easiest place to encounter the Wektu Telu would be at a sacred hot springs near the top of Mount Rinjani, the 3,726-meter-high volcano that soars over Lombok's landscape like the hearth of some divine monarch. And, in fact, the Hindus believe the mountain is the abode of Batari, a goddess who battles evil, while the Wektu Telu know that Nenek, a sexless all-powerful spirit, sits on a cloud-wrapped throne near Rinjani's summit. The Wektu Telu regularly made pilgrimages to those hot springs to pray and make homage to Nenek, and if I made the trek as well, I would likely meet, if not my maker, at least some of the elusive Wektu Telu. I

Opposite. A Hindu Temple on Batu Bolong. Daily offerings are made here to Baruna, the goddess of the sea.

agreed to the plan, and Surya said in three days' time I could make the trek to the sacred hot springs on Rinjani.

In the meantime we drove south, passing the ubiquitous *Cidomos,* jingling two-wheeled taxi horsecarts that employ car tires, and endless emerald rice fields, terraced up into the clouds. An enormous poster of Sylvester Stallone marked the entrance to Ampenan, formerly the main port of Lombok and once a vital link in the spice trade, but now, after being battered by monsoons, little more than a broken-down wooden jetty with rows of deserted warehouses. The road through Ampenan segues imperceptibly into Mataram, since 1958 the provincial capital, and now the island's largest town with some 250,000 residents and a showcase row of square, bland government buildings. The road then turns east, and winds through Cakranegara, or Cakra, the bustling commercial center of Lombok, and until the turn of the century, the royal capital. We stopped at Sukarare, a weaving village where traditional sarongs, *songkets* (the men's traditional shirt), and modern tablecloths were being crafted on wooden backstrap handlooms by mothers and daughters; then at Penujak, a government-run ceramics village where clay pots were being thrown and fired. In both, Surya gave me his spiel, addressing me as "Ladies and Gentlemen," and then subtly steering me into the Lombok equivalent of a Friendship Store, where the locally produced goods could be sold to tourists for many times the real price.

Late in the morning we turned off the paved road and parked next to the vans from three other tour companies. We were at Rambitan, advertised in Bing-Bidy's brochure as a traditional Sasak village, conveniently located just off the main highway, and now a living exhibit for tourists. From a distance Rambitan looked unspoiled, genteel, a village straight from a museum diorama. There were no telephone poles, no aluminum sidings, no television antennas, no indication of a twentieth-century existence. Around the periphery were the traditional rice storage barns, called *alungs,* thatched and hyperboloid-shaped, supported by roughly hewn wooden beams and wide strips of bamboo and banyan tree branches interwoven to form walls and partitions. As we entered the teakwood gate, Surya told me the government had decreed that only traditional-style homes and buildings could be erected in Rambitan, so that foreign tourists, purses brimming with curiosity and currency, could see and experience the old ways. A sort of Williamsburg of the tropics, right down to the traditional costumes and the vulgar selling of a culture. The women who met us as we walked into the village wore black sarongs called *lambungs* covered with short-sleeved overblouses with sharp V-necks. The sarongs were held in place by four-yard-long scarfs, called *sabuks,* trimmed with brightly colored stripes. The *sabuks* were beautiful, and they were for sale. Scores of women and children crowded around us stuffing the *sabuks* in our faces, trying to entice a purchase. Some pushed old square-holed Chinese coins, called *bolongs*; others just asked for money, or pens. It was nearly impossible to walk with the hordes of hawkers pressed against us, and certainly impossible to appreciate the sights. At one point, I ducked into the dark sanctuary of an old man's home, and sat down on a bamboo platform in respite. The old man creaked over to my side, stretched a toothless smile, and pulled something that looked like a fan marked with Sanskrit lettering. It was, in fact, a kind of "Sasak Farmer's Almanac" written on the leaves of the lontar palm, and predicting the weather for a century. I was fascinated, and with Surya as interpreter, enjoyed a session of communication as this village patriarch told me how he had used the almanac in a lifetime of farming. But when I got up to bid goodbye, the elder's smile turned stern, and he asked for money for his time. There was no quaintness left in Rambitan, no natural beneficence towards strangers—only commerce, the oily commerce of tourism. A month previous I had been in Sulawesi, a large Indonesian island to the north, and one day I had been out photographing at the edge of a small village, in an area in the north where tourism was unknown. It was hot, and a farmer passing by noticed my beaded forehead and proceeded to shimmy up the nearest palm tree. He macheted several plump coconut fruits from a hundred feet up, then came back down and split them open for me to drink. The milk was sweet and satisfying, the perfect thirst quencher. I was so thankful, I pulled out my wallet and invited my samaritan to reach in. But he shook his head and grinned, then waved, and wandered off to his field. That could never happen in Rambitan, where every thing, every service, every gesture was for sale. It all seemed to me like some updated version of *puputan.* Only now the invaders were tourists, and the suicide wasn't corporal, but cultural.

Opposite, top. Returning from market with sacks of rice, the staple of all meals in Lombok.
Bottom. A coconut harvest being carried across a papaya field. Because of its rich volcanic soil Lombok is a natural garden, with a cornucopia of foods and spices.

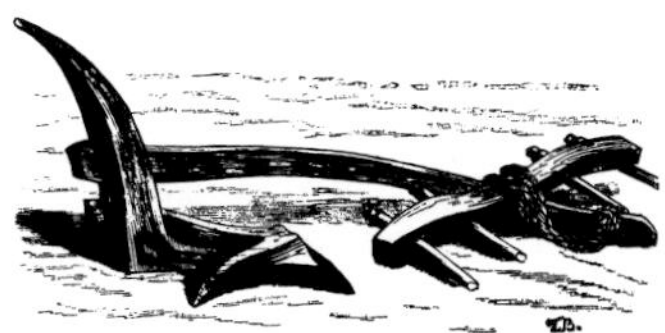

A native wooden plough.

More than ever, I wanted to visit the Wektu Telu. Even though their guiding credo was a corruption of faiths, they promised, I hoped, an honored heritage, a cultural integrity I had yet to find on Lombok. I pushed my way through the clawing crowd, past the other tourists who were engaged in brisk business, crawled into the van, and beseeched the driver to get going.

Lombok means Chili Pepper, and as we continued south in the searing heat the name seemed apt. Not just for the weather, but because experiences so far in Lombok had been sharp and spicy, and lingered long after leaving the table.

Early in the afternoon we arrived at Kuta Beach, on the southern tip of the island. Unlike the infamous Kuta Beach on The Famous Island, there were no topless

Opposite, top. A Wektu Telu praying to Nenek, the god of Mount Rinjani, amidst German tourists at the sacred hot springs near the caldera of the great volcano. *Bottom.* A Lombok fisherman with his monofilament net at Senggigi Beach.

bathers here, no massage vendors, no motorcycles, no surfers, no cold beer. Just a rugged, empty, 60-mile-long coastline, the perfect place to lunch after the suffocating experience of Rambitan. From the back of the van Surya produced a boxed lunch consisting of a hardboiled egg, a piece of chicken, two Kraft processed-cheese sandwiches smeared in butter, and a bottle of Ades, the most popular mineral water in Indonesia.

As I looked around I couldn't help but admire the brutal beauty of Kuta. Limestone cliffs were cragged and pocked, as though beaten by some huge mallet. There were no palm or banana trees, only scrub brush. Except for the small thatched platform that served as our lunch counter, there was no shade. Yet a group of Australian investors, Surya told me, were soon to break ground on a 52-room resort hotel at Tajung Aan, just three miles down the beach. The area, I realized, looked like an Australian coastline, and well it should. As I gazed out over the water to the west I was looking at an invisible demarcation, the Wallace Line.

Though the great nineteenth-century naturalist Sir Alfred Russel Wallace never visited Lombok, he correctly surmised that the island is the westernmost landmass to harbor Australian flora and fauna, making it the dividing zoogeographical line between continental Asia and Oceania. Wallace's Line neatly divides the Indonesian archipelago in two, and is actually the meeting place of continental plates, carrying their isolated cargoes on a collision course. On the one side, the Asian land "barge" has carried wildlife such as apes, tigers, elephants, and rhinos from India and Africa; on the eastern side of the Lombok Strait, no such animals ever roamed, and birds and plants are distinctly Down Under.

That night, between power outages, a troupe of young and supple Hindus performed dances on a stage by the pool. They produced a shower of crystal notes, a spinning kaleidoscopic web of sound in minor keys—the tinkerbells of the *gamelan* orchestra. As six nubile girls in purple *kains* dreamily floated about the stage performing the Gandrung dance, a dance about love and courtship, Surya leaned over to me and confided that he didn't know the route up Rinjani. But, he promised to recruit a veteran guide, and purchase the necessary food and supplies for the three-day trek. It would take a couple days, so in the interim he offered to show me more of the lowlands.

In the next morning's pink dawn Surya knocked on my door. He led me to his van, and my now familiar seat, which was piled with snorkeling gear. We were going to Gili Air.

Of the three tiny reef-rimmed coral islands off the northeastern coast, Gili Air is the closest, just twenty minutes from the village of Pemenang by local *prahu,* a simple outrigger canoe propelled by a long-stemmed outboard motor. *Air* means "water" in Bahasa Indonesia, and the water at Gili Air was beyond belief, clear as mountain air, bursting with the colors of many kinds of fish—butterfly, angel, fire, needle, and goat. Gili Air was a decadently slothful place, where everything moved in slow motion, even the wind in the pandanus and tamarind trees. I lounged around on the silver beach, and periodically slipped into the warm water for a snorkel and a peek at the vibrant coral, and the myriad fish. Time vaporized, and the day drifted away. The island was just 250 acres, small enough to walk its perimeter in just a few hours, but I never moved more than a hundred yards from my towel, and was lavishly content.

On the drive back, we wound up over a 600-meter pass into something called Pusuk ("the Monkey Forest"), a protected state park, and when a troop of grey long-tailed monkeys appeared by the side of the road, Surya turned to me and said, "Ladies and Gentlemen, there are monkeys." I asked Surya why he kept calling me "Ladies and Gentlemen." He revealed that he had taken a three-week course in tourism from the Regional Tourism Office, where the instructors had drilled in his dialogues. They had also taught him how to spot political troublemakers (they often didn't carry a passport) and

Opposite. The Lombokians bathe twice a day, usually in the streams that drain Mount Rinjani.
Right. Young girls collecting coral on the beach.
Bottom. Weaving in the traditional Sasak village of Rambitan.

drug users, both of whom he was mandated to report to the authorities as quickly as possible. After the aborted communist coup of 1965, thousands of suspected communist sympathizers were murdered in Lombok and throughout the young country, relations were severed with China, and a political paranoia set in. It persists to this day, so that many travellers, especially those smacking of idealism, are eyed with a bit of suspicion. I was going to ask Surya about Timor, a nearby former Portuguese outpost that had been seized by Indonesia and was still embroiled in civil war, but decided that the best political science lesson in Indonesia is political silence.

The following morning I took a dawn stroll down what could have been Earth's first beach, down a couple of kilometers to a rock outcropping called Batu Bolong. This was "the sailing stone," a natural archway that juts out into the Lombok Strait, facing the imposing volcano Gunung Agung. I climbed atop the palisade and found a small Hindu Temple festooned with fresh offerings of woven banana leaves filled with flowers, fruit, and chili peppers. These were offerings for Baruna, the goddess of the sea. Legend has it she once demanded the sacrifice of beautiful human virgins who were regularly cast off the Batu Bolong into the frothing waters in an effort to appease her fickle highness. But over the years she seems to have mellowed in her demands. A small boy saw me poking around the temple, and wandered up. When I took a photo of him, he put out his hand for a payment. Setting my camera down, I dug into my pockets for change, and with a smile paid the innocent-looking child. Minutes later, walking down the steps, I discovered a filter for the camera lens missing. I ran back to the temple, but the boy was gone, no sign of the missing filter. Unwittingly, I had made my own sacrifice at Batu Bolong.

A Malay anchor.

Surya spent the morning shopping for the following day's expedition to Rinjani. When he showed up at the hotel around noon I asked if there might be anything to see in the lowlands that related to the Wektu Telu or Mount Rinjani. He ushered me to the van, and off we went to Narmada, about 20 kilometers to the southeast. From a distance Narmada is nothing more than a rise in the road, but once on top of this otherwise ordinary hill an extraordinary sight is presented: a man-made replica of the summit of Mount Rinjani and its crater lake, Segara Anak ("Son of the Sea"). The name Narmada comes from a sacred river in India, and the facsimile of Rinjani was built by a Hindu king of Mataram, Anak Agung Gede Karangasem, in 1805. When he was no longer able to climb the real sacred volcano, the old king, in a variation of Mohammed moving the mountain, had the downscaled copy built. Rumor has it the king was a lecherous old man, and he built a hidden viewing room above the artificial lake. Here he spent his final days leering at the young girls he invited to come to his court for a swim, a sort of Hindu Hugh Hefner with his version of the glass-walled pool. We were visting the grounds on a Saturday, and the lake was packed with young maidens, as well as children, boys, fathers, and mothers out for a cool splash and a picnic on a sunny weekend.

Next Surya brought me to Lingsar ("Running Water"), a large temple complex just a few kilometers north of Narmada. Wektu Telu was once a much more prevalent, popular, and recognized religion, and in 1714 the Lingsar temple was built with the rather rare idea that two religions could co-exist in one house of worship. Thus, Lingsar is designed in two separate sections and levels, with the Hindu Pura on the upper north, and the Wektu Telu basilica on the lower south.

It was late Saturday afternoon when we rolled to the gate of Lingsar. It was chained and locked, with a menacing Mike Tyson charcoaled across a pillar. Surya poked around, and in an adjacent dwelling found the gatekeeper for the temple, an eighty-three-year-old great-grandmother, who listened to our plea to see the temple for higher research purposes. Fine, she said, if we paid her 1,000 rupiah—about sixty cents.

A minute later we were inside the inner sanctum. We looked first at the Hindu

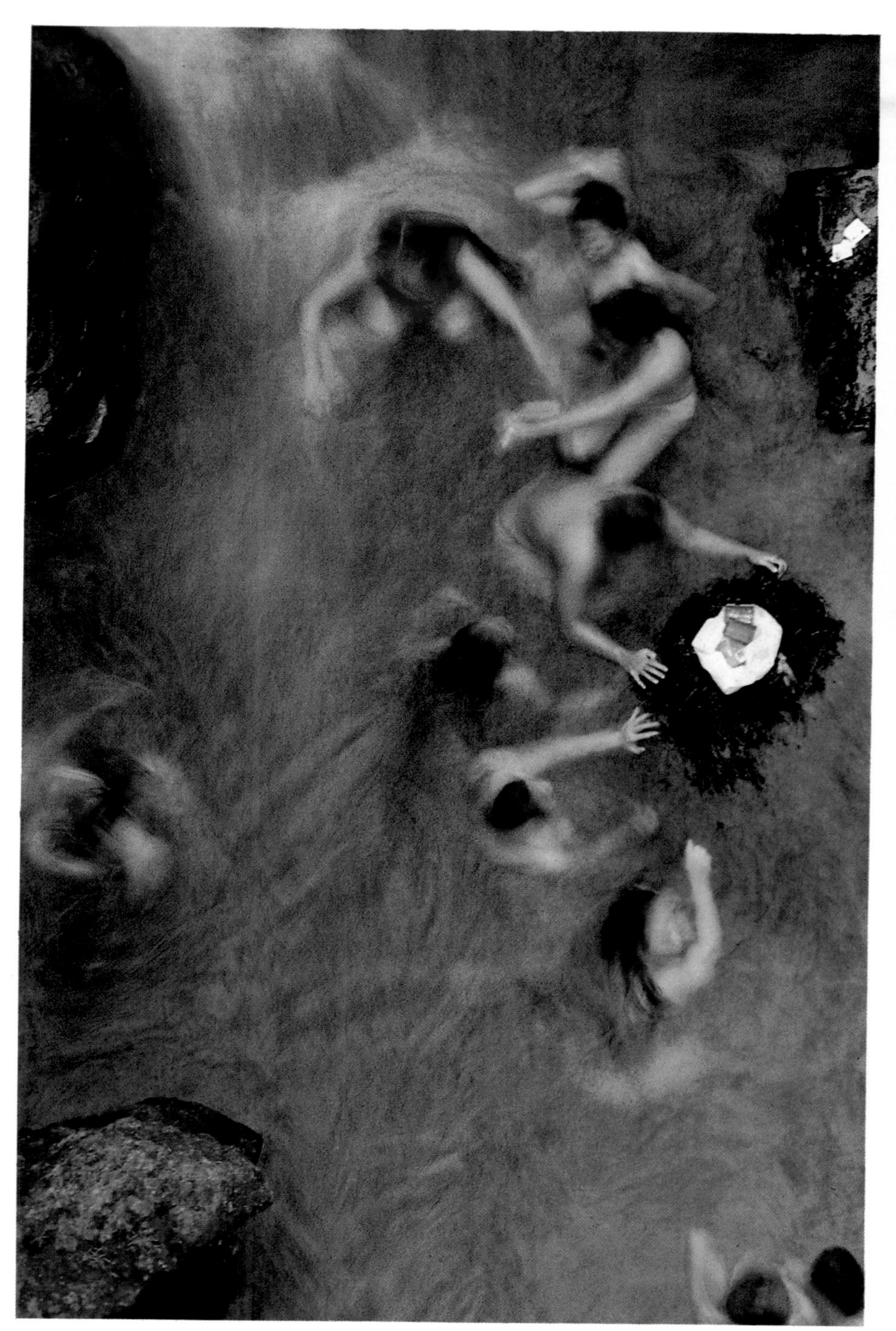

altar, which had a shrine facing Mount Rinjani. Lined in front were a half dozen basalt stones carried down from the summit of the great volcano; they were wrapped in white cloth with yellow ribbons, a tribute to the Rinjani goddess Batari, who always wears those colors.

Opposite, top. Water buffalo tilling a wet rice paddy in Lombok.
Bottom. A boy in a mosque bangs a drum for the call to prayer.

Devoted to Lord Vishnu, the Wektu Telu temple was distinguished for its small enclosed pond in the shade of a Dragon Flower tree. With dainty water lilies floating upon its surface, it looked like a wishing well, and in fact silver and gold coins littered its bottom. Surya started tapping the walls of the little encased spring, and he tossed pieces of hardboiled egg from my lunch box into the clear water. Suddenly a half dozen four-foot-long-plus sluglike creatures writhed out into the middle of the well and started to nip at the pieces. "Holy mackerel!" I exclaimed at the sight. "No, holy eels," Surya corrected. These were the sacred eels the Wektu Telu had worshiped for centuries. Some farmers to this day make their way off the mountain and over to this place to make offerings and feed the eels, who are still a part of their animist cosmology. On the far wall, just beyond the well, was a large-print message in flaring Arabic characters. It was a passage from the Koran, the scripture of Islam, said Surya. For the first time I was witnessing the manifestation of this confusing religion . . . in one holy courtyard all the bases were covered, with tributes to the Hindu god Vishnu, the animist eel gods, and Allah. I covered my own bases by tossing three coins into the fountain, and making a silent wish for good weather for the next day's climb. With that we returned to the Senggigi.

In my predawn dusty consciousness I heard the knock. I checked my watch. It was 5:00 A.M., time to go. Benjamin Noya, a freelance Rinjani guide, met me at the door, and urged my haste, as it would be a three-hour drive to the starting point for the 30-kilometer trek. In minutes I tossed my pack into the faded green van and sat back to enjoy the sunrise ride. Surya was not with us—he had made it clear he didn't cotton to mountain climbing—but in his stead was Ketut Karune, a trainee guide for Bing-Bidy who boasted that he prepared for this trek by running five kilometers on the beach the previous Thursday.

A racquet-tailed kingfisher.

On the ride north, past the black sand beaches that are testimony to Rinjani's last eruption, in 1901, Ben revealed a bit of his background. He was twenty-six, a Christian (an oddity on Lombok), and one of nine children born to Protestant parents from the Moluccas, the Spice Islands to the northeast. He had made fifteen successful trips up Rinjani between his continuing studies in economics at the university. He saw trekking tours as an up-and-coming business opportunity, and was saving his rupiah so he could start his own concern, Rinjani Trekking and Tours. It would be a seasonal affair, as during the monsoons climbing Rinjani is dangerous and extremely difficult. For our trek, in mid-April, the rains were just ending, and this would be Ben's first tour in eight months.

Around 8:30 we rattled up to the end of the road somewhere around six hundred meters above sea level, at Senaro, another traditional Sasak compound consisting of about twenty thatched huts enclosed by a wooden paling fence. The guidebook claimed it had been just a decade since the villagers of Senaro saw their first Westerners, but it took less than a minute for an elder to corner me and present a piece of stationery from a Mataram hotel with a handwritten message in perfect English, and excellent penmanship as well: "We kindly request you make a donation to our village so that we might save up enough money to buy supplies to make weavings, which we can sell to tourists to make enough money so we can pay for a doctor to come and make regular visits." The old man gave me a greedy smile, permanently beet red from a lifetime of chewing betel nut (a mild narcotic that is the drug of choice throughout Asia), and extended his hand. It was a persuasive approach, certainly better than the seedy solicitings of Rambitan. I found myself reaching for a handful of rupiah to make my contribution, though I wasn't totally convinced the money would be spent as the letter promised.

As Ketut and I unloaded the van, Ben went off to find porters. Apparently Swastika Travel had been here the day before, and had hired off the first string of porters. So Ben had to go searching in some of the backeddy villages. The plan was that Ketut and I would head up the trail to a lunch-stop shelter, and wait there for Ben and his men.

The hike was up a wide path through a dense forest of mahogany and teak, trees alive with the cascading calls of parrots and lorikeets. An hour into the ascent we stopped in our tracks at the grunt of a wild pig, then looked up to a bolt of white wheeling through the branches. It was, Ketut guessed, a sulphur-crested cockatoo, a bird of Australian origin whose farthest roost west is Lombok, on the edge of the Wallace Line.

A couple of hours into the climb, the clouds turned the color of a bad bruise and closed in. Then the heavens cut loose, sending down thick columns of rain, turning the trail into a mudslide, mocking my offerings at Lingsar the day before. Battered and cold to the bone, we slogged along, finally reaching the little shelter around noon. We pulled off our soaked clothes to discover leeches with their bloated, rubbery heads buried into our legs. As we picked them off we shivered violently in the storm. This was not fun, and as I slapped my sides for warmth I wondered about the wisdom of this little exercise. Two hours later, Ben trudged up the trail, then climbed onto the platform with a worried look on his face. It seemed that after an extensive search he had at last recruited four porters. But halfway up the trail, one sprained his ankle and had to head back, and another decided he didn't want to attempt the three-day trek as the following morning was the beginning of Ramadan, the annual month-long Islamic fast that requires the faithful to abstain from food and water from dawn until dusk. I had been warned not to visit Lombok during Ramadan, the caveat being that the people were cranky and restaurants closed during the fasting hours. The admonition seemed silly to me at the time, but now it made more sense. Ben said he had abandoned our packs down the trail, and had come to tell us the news before heading back down. He confessed concern about finding additional porters: nobody wanted to carry heavy loads up the mountain if he couldn't eat or drink.

There was little we could do save continue, carrying as much as we could, and trusting Ben could recruit the needed personnel for the rest. Ben demonstrated little sophistication in the packing for the expedition. He had thrown in two cases of Ades, a dozen heavy tins of meat and vegetables, a jerry can of kerosene, a large stove, various pots and pans, bags of rice, a lantern, and other paraphernalia; so much stuff that it took two porters for every client. I promised if he ever got his trekking company off the ground I would send him a supply of freeze-dried food and a camping stove.

When the rain finally ceased, Ben headed back down the trail, while Ketut and I plowed upwards through the muck. Ben was hopeful he could collect the new porters before sunset, and reconnect with us by midnight. Ketut, who had never been in this part of Lombok, was uneasy trekking through the forest with such a small party, and he stopped to strap a knife on his ankle, just in case we came across any unfriendly Wektu Telu. At dusk we reached the base camp hut near a small artesian spring. We watched a spectacular sunset across Lombok Strait, then collapsed in sleep, hoping to wake up to Ben and our supplies.

It didn't happen. With daybreak, there was no sign of Ben. So, after a brief breakfast of cold, glutinous rice, the last of the food we were carrying, we continued upwards into the mist, stepping up what I now believed was our private stairway to hell.

That all changed when we reached the rim of the caldera, at a vista called Pelawangan, somewhere around 2,600 meters, just above the sailing clouds that snagged on peaks below us. We looked down at a sublime sight, a replica of the scene we'd seen at Narmada, only this time it was the real thing: the revered Segara Anak, the algae-green holy lake. It was much more magnificent than I had imagined. Rimmed with pine trees, backdropped by the pyramid summit of Rinjani itself, and punctuated with the black seamed and pleated cinder cone of Gunung Baru, it looked like a spent reactor cooling tower at Three-Mile Island. Segara Anak itself reminded me of Lake Tahoe seen from Heavenly Valley, but before the casinos and condos. It possessed a primeval beauty, awe-inspiring in its grandeur, and it was easy to see why the Hindus and the Wektu Telu had decided this was the throne-room of their mountain gods.

It took three hours to pick our way down the steep inside slopes of the caldera, down to the cold waters of the lake, then around its perimeter to an exit stream (Kokoq Puteq, which means "White River") just beneath the leaning saw-toothed crown of Rinjani. I had been told no fish swam in Segara Anak, but at the source of this stream the waters were churning with activity, and upon inspection I saw what looked to be dozens of trout dancing in the currents. The sight only reminded me of how hungry I was. Ben still hadn't showed, and he had all the food. Here it was the first day of Ramadan, and I was fasting, probably the only Christian on the island to be doing so.

Following the stream a short ways we came to a cliff, over which the water burst in a flash of foam and white. Sixty feet below was a level area spotted with steaming pools. These were the sacred hot springs, our goal. There were people milling about, perhaps the Wektu Telu, and I scrambled down to meet the souls of my quest. They weren't there. Instead, there were European tourists, Swedes, Dutch, Swiss, and German, all here to soak in the scenery and the hot springs. We set up our camp between the Dutch and the Swedes, who were in the middle of a clamorous card game, and set out for a bath while we waited for Ben and any Wektu Telu who might show. I found a pool next to the raging stream, just beneath the falls, so I could control the temperature by simply floating towards or away from the cold creek. The soak was celestial. I laid back in ecstasy, watching a troop of black monkeys across the stream watching me. An hour passed, I think—

Below. Most fishing in Lombok is at night during the nine-month dry season. Each boat is painted to its own personality, to please Baruna, the goddess of the sea.

In the nineteenth century, native workmen of Lombok made their own guns. They used some parts taken from English muskets, but most of the necessary pieces were crafted with a mud forge, a bamboo-and-feathers bellows, and a few files and hammers. The gun barrels were bored with a bamboo basket full of stones, shown here. The final product was a well-finished, true-shooting rifle.

time warped in the springs. But at some point Ben and his porters, four Muslims, came over the rise, and I rousted myself out for a much-needed meal, something Ben whipped up that consisted of eggs, asparagus, crab, and hot dogs. It tasted surprisingly good.

It was sometime before twilight that two dark figures quietly walked into our camp and over to the edge of the river. It was a father and son team, dusky-skinned with straight black hair and barefoot. They carried no backpacks, no accoutrements, save a cloth bundle carried like a hobo. Ben walked over to my tent and whispered, "Wektu Telu."

I watched as the father and his boy unpacked their prayer rugs and unfurled them on a rock between the foaming river and the springs. Then, as a pair of naked Germans splashed and giggled and soaped up obliviously in the foreground, the Wektu Telu went into a deep meditative trance, called *Bertapa,* and prayed to Nenek, the god of Mount Rinjani.

I asked Ben if we could speak to the Wektu Telu. He advised against it for a couple of reasons. First, Ben, being a Christian, did not speak their language, and they certainly did not speak English, nor Bahasa, the official Indonesian language. Second, if asked, the Wektu Telu would deny their religion and claim they were simply Muslim. Though they believed in magic and spirits beyond the conventional gods of Islam and Hinduism, they had to hide those extra theologies from the outside. Ben explained that though officially the government allowed any religion, in reality a discrimination campaign was being waged, an effort designed to turn the remaining Wektu Telu into the proper Muslims that comprise 90% of Indonesia. The Wektu Telu had no voting privileges; it was difficult for them to buy land. The Muslim farmers in the area were paid better prices for their crops. The message was clear—life would be better for the Wektu Telu if they would convert.

Yet some never would. Their beliefs, however incongruous and queer, were stronger than any other force in their lives. A compromise was unthinkable; they lived with an integrity that seemed gone from the other Lombokians I had encountered, gone from my world. They believed in fire-emitting magic swords and flying white horses that rode to the top of Rinjani, but they didn't believe in the almighty rupiah, or tourism. As I watched the Wektu Telu go through their benedictions, I was impressed, perhaps for the wrong reasons. They never acknowledged the existence of any of the polyglot tourists who were using their sacred place of worship for a hot-tub party. They never complained, never intruded. They didn't scrawl graffiti on the rocks. They didn't drop any litter. They never tried to sell us anything. And, they never asked for money. They prayed, they spent the night wrapped in thin blankets, and as quietly as they came, they left.

The following evening I was back off the mountain, tired and tempered from the Rinjani experience. It was a sultry night, one that would have quickened the blood of Mel and Sigourney. I wandered across the street from the hotel and found a little *warung* (restaurant) called Pondok Senggigi ("House of Senggigi"). It was packed with young Western tourists wearing sarongs and bright billowy shirts. They drank beer while a stereo tape-deck alternately played James Taylor and *gamelan.* I looked at the menu, and was struck by the prices—they were a tenth those across the street at the Senggigi Beach Hotel—and by the offerings. They ran from traditional dishes, such as Lombok spinach with hot *lombok* (chili) sauce, to cheeseburgers and fries; from *lassis* (liquid yogurt) to papaya milkshakes; from *nasi goring* (traditional fried rice) to freshly trapped lobster. It was confusing, as though the place didn't know what or where it wanted to be, East or West, local or foreign. As I sat down, a waiter, perhaps a future tour guide, walked over to my table in the characteristically small steps of many Indonesians. He was wearing a T-shirt that read "Where the Hell is Lombok?" Nobody, I thought, not even the Lombokians, really knows.

Honshu

Pedaling the Path of Small Knowing

This is God's Food,
Don't Eat it!
—SIGN AT THE KIYOMIZU TEMPLE, KYOTO

It is a land as intangible and intoxicating as perfume, a paradox as large and implacable as the Sphinx, and I hoped to explore it on the back of a bike. The only problem was the last time I rode a bike it had a little bell on its handlebars.

Now, thirty years later, I was mounting the chrome-moly steel frame of something named Hard Rock. A mountain bike with fat tires and a low center of gravity, it was not too different in appearance from the Schwinn I rode to Wood Acres Elementary School, except this version had seventeen additional gears. And I would need every one and more to negotiate this particular landscape.

I was at the top of a hill in a south-central district called Kinki, a prefecture called Wakayama, on the island of Honshu—perhaps, to the Westerner, the most enigmatic acreage in the world. While the island itself may not be familiar, some of its landmarks are: Tokyo, Osaka, Kyoto. Yet it is much more than the berth for some of the world's most celebrated cities; it is an island of rugged mountains and rip-roaring rivers, and vast expanses of dense forest. It is an island of oxymorons: samurai and monk, chrysanthemum and sword, iron lanterns and neon, an ineffable aesthetic refinement coupled with a lust for Madonna memorabilia. It is a culture that declines to be known, a place which makes sense to few of a Western consciousness.

I wanted to see if I could claim just a sliver of understanding, take a peek behind the fan, with a passage through Honshu. I figured an ideal way would be by bicycle. Cars were for cities and freeways; walking for wilderness. I wanted to explore the in-betwixt, the central Kabuki principle of *ma,* the space and/or time between, and the vehicle of choice itself defined a middle ground. So I signed with a company called Travent Tours of Waterbury Center, Vermont, an organization specializing in overseas bicycle trips and featuring a unique passage through Japan. It was led by Tada Yasue, forty-five, a short, oaken Honshu native with a conspiratorial grin and legs like piano wire. He is perhaps the preeminent backcountry biker in Japan.

As I pulled on my newly purchased Novara gloves and adjusted my Bell Ultra-Light tortoise-shell-shaped helmet I looked around. I could not quite believe this was Japan. We were 800 meters above sea level on a mountaintop called Tentsuji, thirty-five kilometers above the city of Gojo. Looking down on all sides we could see nothing but

Opposite. Cruising down the Hiki Highway in southern Honshu.

forest: maples, pines, incense cedars, spruce, and other evergreens. The area is called the Tibet of Japan. I had never imagined an island as populated and crowded, with million-dollar-a-square-meter properties, with twice as many television sets as people, could offer so much wilderness. It was exquisite.

Patrick Delaney, the assistant leader, a twenty-four-year-old cherub-faced Coloradan, noticed I was fumbling with the equipment and came over to assist. "Know how to shift gears?" he asked. "Shift, no," I smiled. He gave me a crash course, playing the two levers like piano keys, telling me that the ideal rate was to have my legs moving at sixty to one hundred rotations a minute whatever the grade.

Then all twelve of us donned our helmets, hopped on our horses, and headed downhill. This dirty dozen looked more like a parade of teenage mutant ninja turtles off to the pizza parlor.

Our route would take us from the heights of the Yoshino mountains of southern Honshu to the Pacific, literally a journey from summit to sea.

The first few miles were easy enough as I learned how to use the brakes (pump lightly every few seconds, so as not to burn the rubber) and tentatively played with the shift levers as we hit a level spot or a small incline. Tada, our elfish guide and *sensei,* had given us each a photocopied sheet of fourteen photos of landmarks along the way, signs in *kanji* (Japanese characters), buildings, bridges, tunnels, so that we could keep track of our route. While he rode in front of the pack, Patrick brought up the rear, and each was in contact with the other with walkie talkies. It seemed fail-safe.

Ah so, not so. In the first two hours of riding we somehow lost Valerie, an energetic redhead who hosts a popular midday talk show on Canadian television. Her husband,

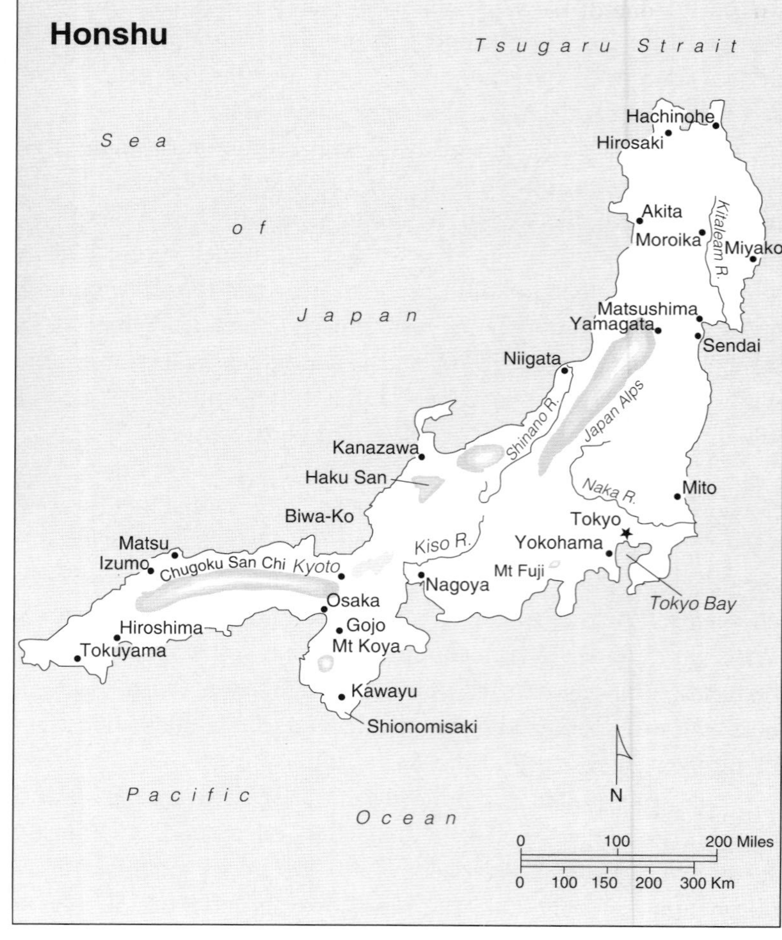

Andrew, panicked when Patrick, who had been in the rear, pedaled up to the vending machine where a few of us had stopped for refreshments. Valerie had been between Andrew and Patrick, and hadn't been seen for several kilometers. While Patrick and Andrew headed back to find our first lost soul, I lingered at the vending machines on this lonely highway, and widened my eyes at the contents. Not only soft drinks and hot drinks of every kind, including many I'd never dreamed of (honey-pomegranate; bean-yogurt; something called Pocari Sweat; a vegetable drink called Toughman), but also beer: Sapporo, Kirin, Asahi Super Dry, and the like, in all sizes, from pint to liter and beyond. What kept kids from using these machines, I wondered? How could they offer this on a precipitous mountain highway? It was a puzzle that perplexed as I saddled up and headed down the trail.

Opposite. A brass Buddha in the high forests of the Yoshino Kumano National Park in the Kinki district. *Below.* Nachi no Taki, the highest waterfall in Japan. Jimmu, the country's mythical first emperor, supposedly worshipped at these falls.

Minutes later Valerie was found. She had correctly followed a route marked "detour," while the rest of the party had missed or ignored the sign.

We continued the downward trek, down a road like a logchute, through the Yoshino Kumano National Park alongside the Nyu River. I realized I liked this mode of travel: the crackling landscapes, the delicious wind in my face, the smell of pine tar and creosote packing my nostrils, the snapping sound of the triangular flag attached to my rear fender, the sound of speed. There was an inner calmness to it all, as though lounging in the eye of a hurricane—exhilarated yet safe, powerful yet serene, feral yet tame.

Then someone yelled "Car back," cyclist slang for "Watch out, a car is approaching from behind." I tensed up; pulled as close to the left curb as I could as a Mitsubishi van whizzed by, missing my pedal by inches. It brought me back to reality, and reminded me of the flip side of this activity, and this land—danger and violence. The island whose side we were spinning down, though even-tempered and peaceful on the surface, is one of the most seismically active in the world. It floats atop the eruptive Pa-

Opposite, top. Dawn in a prefecture called Wakayama, the most enigmatic acreage in the world.
Bottom. The Yoshino mountains, the Tibet of Japan.

cific Plate. And the people, as genteel and shy as anywhere—the babies don't even seem to cry—have a history of blood and brutality as extreme as any in the world. Minutes later, as if to illustrate this thought, Jan, the most experienced cyclist in our group, wiped out while steering onto a sidewalk, nastily scraping his elbow, knee, and thigh. He smiled through his pain, as any Japanese would, and kept on going.

We next stopped at Tanise no Tsuribashi, where the longest pedestrian suspension bridge in the world sways over the Nyu, and pulled out various snacks. The Westerners munched on trail mix with Reeses Pieces and peanut butter and crackers, while Tada crunched on a dried yam. Japanese tourists, here to see the famous bridge, turned their cameras on us. As we snacked, dozens of cameras clicked furiously, capturing our every bite and movement. For the first time in my life I felt like an animal in a zoo, or like the hundreds of people I've photographed in third-world countries over the years—the people in exotic dress and dark skin I've brazenly approached, pointing lenses in their faces. Now I felt like stepping to the Japanese tourists and asking for yen.

Butsudan is a Buddhist household shrine containing family ancestral mortuary tablets (*ihai*).

The journey continued, as we coasted down a quiet Route 168, as scenic as any stretch of the California coast's Route 1. The ride was graced, almost transfigured, by a beauty that touched the heart. It could be called wilderness, except for the step dams that sliced up the river, the tunnels that bored through cliffs, the convex mirrors at every bend in the road, and the never-ending vending machines. At one I sampled a Suntory all-malt beer "For a Refreshing Day," its label promised, and noticed the ground was littered with aluminum pop-tops, like American beaches just a few years ago. In a land with such a sensibility of discipline and order I was startled to see this. Soft-drink cans around the world now had the pop-tops that stayed attached to the cans, and yet Japan, with all its aesthetic sophistication, still utilized an out-of-date system that littered the ground and threatened the landscape. It was an unexpected ragged edge.

Toward the end of the day the chain on the bike ridden by Pamela, the photographer for the expedition, broke, and the convoy halted for review. Pamela, who had once spent nine months cycling from London to Beirut, remarked that the scenery was unparalleled, unlike any she'd ever seen. She didn't want to miss a minute of biking. The repair would take some time, so Tada and Patrick each offered Pamela his bike. One would sit out the rest of the day in the support vehicle, a white Nissan pickup. (White is the color of 90% of the vehicles in Japan.) Pamela surveyed the options. Tada's was a four-year-old Japanese model held together by bailing wire that looked five times its age, while Patrick's was a shiny new touring model. In minutes Pamela was sailing down the road, while Patrick bounced behind in the pale truck.

The last stop of the day was Kumano Hongu. This ancient Shinto shrine was once on an island in the middle of the river, but it had been flooded so many times it was moved, piece by piece, to a place on the high ground above the banks. Inside we watched a parade of monks in black and gold robes padding by. They moved within a realm of rarefied abstraction, reserved and in control, but I knew that this might be appearance as much as anything in Japan. Just a few days previous while in Kyoto we had cycled the cherry-lined Philosopher's Walk to one of the city's 1,700 temples, a classic Zen temple called Nanzen-ji, built in 1921. While wandering through the complex Pamela looked for a bathroom but could not find one. So Tada poked his head into a monk's *hojo* and asked if Pamela might use the facilities. At first the monk said no, but when he saw Pamela and her obvious discomfort, he nodded for her to come inside. As she stepped up the stairs the monk threw a small fit because in her haste she had neglected to take off her sneakers. The monk forced her to backtrack and start again. While Pamela was down the hall I walked into the monk's sparse main room, admiring his hanging scrolls and arranged flowers.

つりぼり場

When he returned through a sliding door I gave him a broad smile. He looked back with a deep sadness, and then started to speak in the most fractured of English about his wife. She had suffered an aneurism six days ago, and just this morning had eaten her first food. When Pamela returned to thank the monk, his face suddenly went glassy. He collapsed in my arms and began to sob uncontrollably. I hugged him, patted his back, and tried to comfort him, but for several minutes he clung to me, desperate for human contact. Pamela took the monk's hand and explained that two of her friends had suffered aneurisms and had made complete recoveries. He finally regained self-control, bowed low several times, and thanked us both for our understanding. He was still wiping tears from reddened eyes as we departed. I had thought public emotion was heresy for a Japanese; for a Japanese Buddhist monk it had seemed unfathomable. But it demonstrated the very human side of a man who at his core was like any other on the planet, and it touched me deeply.

After touring Kumano Hongu we cycled another five kilometers before pulling into home for the evening, the Kinokuni Hotel in the spa town of Kawayu. We had cycled ninety kilometers, almost sixty miles, more than the sum total previously pedaled in my life, and I could feel the tally. We checked in at the desk and were told through Tada (nobody spoke English in this part of Japan) that we didn't need keys, as theft in the countryside was unheard of. Nonetheless, when I slid back the door to my tatami-lined room, I was met with a low, lacquered table, a television set, and a thick, grey safe. Preparing for the future, when more *gaijin* (outside persons) would visit, I had to assume. I turned on the television to a classic American game, transported and transliterated to Osaka. Japan's favorite team, the Yomiuri Giants, was playing its traditional rivals, the Hanshin Tigers, and the stadium was packed. Despite the familiar trappings, I couldn't follow the game, and my legs were beginning to feel the day, so I changed from my Lycra into a floppy-sleeved *yukata* (a printed cotton kimono), slipped into a pair of the requisite leather/tan one-tiny-size-fits-all Japanese slippers, and wandered downstairs. I found my way to the *onsen* (hot springs) at the edge of the river for a much-needed soak and some *sake.*

This was the third hot soak of the trip, and the first co-ed affair. It was impossible to be modest, as the towels provided (*tadru*) were no bigger than washcloths. Our group consisted of some conservative souls, several veterans of cycle trips through France and Italy where each evening was spent in elegant, private-quartered chateaus, and now the concept of co-ed bathing wasn't overly appealing. Where else would a group of well-sneakered strangers suddenly strip and share a bath but on a cycle trip through rural Japan? I saw it as a moment of unalloyed egalitarianism, and agreed with the Japanese saying, *Hadaka to hadaka no tsukiai*—"Relationships between naked people are honest relationships." Our first soak, in Kyoto two days previous, had proved such. After a day of visiting temples, we parked our bikes at the *ryokan,* a former brothel, and walked down the rectilinear street to the *ofuro,* the public baths. It was a segregated affair, women on one side, men the other. But surprises were within. After sampling a couple of the five baths, I eased into a deep, well-liked tub, something called *harinoyu* or "needle bath," and immediately felt like I'd put my finger in a wall socket. Electric current was racing through the tub. I forced myself to submerge my head, but I couldn't take the jolts for more than a second. I jumped out of the tub, still shaking from the shocks. "What is that?" I asked the room, but nobody spoke English. Some sort of liquid electrolysis, I figured, and joked to myself that it might explain the lack of body hair on the Japanese.

What they lacked in body hair, however, they made up in tattoos. Spread across the bodies from neck to calf of every Japanese in the bath were the tell-tale marks of the Yakuza, the Japanese mafia. These were no ordinary tattoos, but magnificent, full-color designs of samurai warriors, flowers, sake gourds, and dragons. We were, as it turned out, in the heart of the third largest Yakuza neighborhood in the country. It was a bit discon-

Opposite, top. One of the many riverside hot springs in the spa town of Kawayu. *Bottom left.* A monk at daybreak meditation at the Seiganto-ji Temple. *Bottom right.* An aesthetic perfection prevails in the temple compounds.

Shoryobune are little straw "boats of the blessed souls" with sails inscribed with posthumous names of the dead. These are launched during the Bon festival that honors the spirits of ancestors.

Opposite, top. Inside the Higashi Hongan Ji in Kyoto. The temple, one of the largest in Japan, is the seat of the dissident Jodo Shin shu sect.
Bottom left. Along the Philosophers' Pathway around Ginkakuji in old Kyoto.
Bottom right. A Shinto priest at the Kasuga-taisha Shrine in the ancient capital of Nara. The city witnessed the introduction of Buddhism into Japan.

certing at first, but everyone seemed at peace in the baths, and I went about my business. Then, while squatting on one of the plastic stools and pouring hot water over my head with a plastic bowl, shampoo ran into my eyes, and I dropped my Lifebuoy. I groped around for it with eyes pinched shut, and then someone slipped it into my hands. When the sting abated I opened my eyes to thank whoever had helped me. A large, older man with translucent skin was squatting in the stool next to me, while a young man with a shaved skull performed a rigorous massage on his back. The old man turned to me, smiled, and nodded. I couldn't help but notice he had the most elaborate tattoo of all, and that the little finger on his right hand was missing.

Still suffering a bit from jet lag, and my neck a bit kinked from sleeping on the Japanese husk-filled pillow, I awoke early in Kawayu. The river outside my window was spinning its currents with a sound like a wind chime. Everyone else was still asleep, so I wrapped my striped *yukata,* and wandered downstairs to the breakfast room. It was 5:45, and service wouldn't start until 6:00, but the attendant allowed me in just the same and served me green tea. I sat back and sipped while perusing my guidebook, searching in vain for some reference to this area. Hearing a chorus of voices, I looked up. In a corner were all the hotel attendants and employees, gathered in a circle and chanting. The leader would read something from a small book, and then the group would cheer in unison. It was like a pregame psych-up, and it went on for fifteen minutes. Then, with one last shout, the group split and members busily went their way. I was instantly served another cup of tea, then a raw egg, miso soup, pressed seaweed, dried fish, pickled plum, tofu, and boiled rice. I fiddled with most of the food, picking at it with my chopsticks, and managed to get some down, but I couldn't stomach the raw egg. When few were watching, I snuck back to the kitchen and asked the cook, with the appropriate gestures, if he could give it a fry. He nodded gracefully several times, disappeared, and minutes later reappeared with a greasy but definitely fried rendition, which made me happier for the moment—until two whey-faced women attendants padded over to my side and began giggling in sibilant voices. I thought they were amused at my eating habits, until one tugged at my *yukata* and giggled more. She made a turning motion with her hand, and I understood. I had my *yukata* on inside out.

Samurai giving instructions to soldiers.

The day was a layover, a chance to explore an area few Westerners ever see. I rode my bike around a bit, and wandered around the village, up and down the river. When I reentered our hotel I noticed a poster on the stripped-oak wall, a wanted poster issued by the police. I glanced at a photo on the poster, then did a double take. The picture closely resembled the man who had squatted next to me in the Kyoto hot baths two days ago, the man with the missing finger who had handed me the soap.

The next morning there was no rising sun. Instead the day broke foggy and wet, the smell of woodsmoke and rain drifting in the open window. It was supposed to be our first attempt at a serious 19-kilometer uphill ride for the first hours of the day, but the rain squelched that plan, and I was privately happy for it. Instead, Tada arranged a bus, and we all rode in comfort to the top of the Kobiro Pass to a brand-new blond-wood Western-style chalet, a *dojo* built by a friend of Tada's. Mr. Hino was a martial arts expert. While we waited for the rain to withdraw Mr. Hino demonstrated his private-label brand of martial arts, something called *Shiyoken,* a technique based on the central notion of *ki,* or spirit power, which, like the guiding principle of judo, uses the energy of an attacker against himself. Perhaps because I was the largest *gaijin* in the group, Mr. Hino asked me to attack him. I stood in the center of the room feeling like a gross, big-nosed brute towering over the slight, almost fragile-looking Mr. Hino. Then, with the signal, in sinister fashion I snarled and grabbed Mr. Hino's arm in an attack. Within nano-seconds he had me off-balance, ready to fall over backwards and cry uncle. He had absorbed my energy into his

主神社
愛
縁祈願・恋人祈願
賽銭

Opposite, top. The Seiganto-ji Temple, overlooking Nachi falls. This is the starting point for pilgrimages to the thirty-three Holy Kannon Temples.
Bottom. The flowers of a private garden along the Kii Peninsula.

own, made my strength his own. It occurred to me that his method was not unlike his homeland in its absorption of nearly every Western technique—beating the West at its own games of technology, trade, real estate, and eco-banditry. In the near half-century since the war all the quiet, catlike moves of Japan had gathered an imposing weight, and now the scales had tipped eastward. Today some argue that the Japanese never lost the war, but simply employed the spirit of *ki,* and have been patiently waiting for the true and final score.

By early afternoon the rain was gone and everything was mute. The distant mountains were veiled behind a screen of clouds that sometimes thickened, sometimes parted, sometimes drifted across the tops to register the changing moods of the landscape. As we clambered aboard our bikes the sun peeked through, and we pointed our panniers towards the sea. We barrelled down Highway 311, the ancient road the emperors had travelled to visit the hot springs. Once again the ride was almost all downhill, so I never had to pedal to prove my mettle, just perfect my braking technique, for which I discovered I possessed a natural talent. There was just one upgrade during the day, a slight 10% one, and there I fell behind the rest of the group, using the excuse that I wanted to take some photos. But even in my solitude I became embarrassed when two prepubescent boys with one-speed *jitensha* (Japanese bikes) came whizzing by me with universally understood grins.

Kannon, the goddess who rules over paradise, is often pictured on a carp.

The afternoon was the most glorious day of cycling of my existence. Never mind that I could count less than half a dozen. It was the most spectacular road I had ever travelled, winding through cathedrals of cypress and chestnut, groves of bamboo, by whiffs of orange blossoms, with the roiling and quite inviting Hiki River beneath us. The river leapt with rapids for long stretches, then purled into eddies, pools of quiet water that when painted with the afternoon sun looked polished with the impossible finish of a lovely lacquered screen. The surface of the river at these spots was as exquisite as it was opaque, much like the land through which we passed.

The region reminded me of the Sitka National Forest in Alaska, as pristine and lyrical as any I had ever seen. With but a few small exceptions, the hills were carpeted with lush first-growth forest. It seemed a place where Nature had held her breath. But while basking in the aesthetic clarity of the place I couldn't help but wonder how the Japanese could ably preserve such a beautiful spot on their most populated island, and at the same time act so criminally with forests overseas. As the world's largest importer of wood, Japan has had to create enormous underwater holding pens off its coastline to warehouse its massive purchases. And because the Japanese are able to buy wood from the U.S. at such low prices, there has been no cutting of their own forests for the last fifty years. So, while the island nation has omnivorously cannibalized the world outside, it has maintained something of its own environmental integrity.

The end of the day brought us to the rolling Pacific, and we pedaled the last few kilometers along an old cement strand, passing what appeared to be bunkers left from the war. We were on the southern stretch of the Honshu coast, and Okinawa wasn't all that far away. If the war had continued, this might have been a beachhead for the Allied forces.

The Kokuminshukusha Furusato Hotel was a run-down affair with cracks in the cement walls and water stains creeping across the ceilings. Still, it had Western-style toilets, which had a few in our group excited. The stalls even had graphic instructions on how to use them for the unfamiliar Japanese. Nonetheless, I was surprised when I finally sat down on one, and I jumped back up in a start. It was a hot seat, and I feared another electrolysis encounter, but looking closely I saw the cover was plugged into the wall and a little heater was attached. I eased myself back down, and found the heated seat quite comfortable. I wondered why such a wonderful device had never made it to *my* hometown.

The next day we rode southeast, along the Karekinada coast road, Route 42, towards the southern tip of the Kii Peninsula. The highway stitched across coves and little bays, passing fishing villages and vending machines. At one overlook we stopped to admire a string of giant paper streamers shaped like *koi* (fish) flapping over a surfers' beach. They were leftovers from Kodomo-No-Hi, Children's Day, a national holiday in which the characters of the carp, strength and determination in its upstream swim of life, are celebrated as hopeful qualities in Japanese youth. Down the road, for the first time we passed warrens of roadside homes: low, light, wooden structures with brightly colored roofs, their first stories open to the street, and thin awnings sloping back to miniature balconies on paper-screened second stories.

It was mostly level cycling, with the occasional modest slope, which didn't pose much of a problem, and it allowed me to experiment with different gears. The day was humid. The light had a strange, oblique intensity that gave the faces we passed a tranced look, and the landscapes the appearance of being immobilized under glass. At lunch Pamela and I parked at a restaurant on a cliff overlooking the ocean, a spot called Lover's Cape. A sign explained a phenomenon in which waves from two directions met at this point, coupling like lovers, creating something called *meotonami* (the wave of a married couple). I watched the waves for a couple of minutes, and thought that nothing better described the Japan experience—two opposite waves of experiences and values crashing together, creating a single operating body that works. There was also a sign pointing southeast that stated we were 10,000 kilometers from San Francisco, my home. It reminded me of what I had recently read: that in the Middle Ages, when Japanese priests reached a certain age and rank, they were sealed inside small boats and set adrift in the currents off the Kii Peninsula. They believed they would reach paradise, somewhere off in the direction of California.

Crests or badges were the expressive form of Japanese heraldry. The family symbol, or *mon*, was known in Japan as early as 900 A.D. and reached its highest development during feudal times. The *mon* was used on everything that belonged to a family.

Late afternoon I stopped at another of the ubiquitous vending machines. I was amazed to encounter one offering hard-core pornographic magazines, again right on the road, for anyone to purchase. (Even correct change wasn't required, as the machine would change most any paper yen combination.) What to make of a people with such an exquisite gift for purity as well as an unrivaled capacity for perversity? How far could the stoic hedonists go?

We finished the day with a six kilometer uphill climb, the hardest yet. It led to a plateau, Cape Shionomisaki, the spit of land that marks the farthest point south of the island, the nubbin on the bottom of Honshu. Next to the lighthouse was a lodge called Nishida, where we unwound while Tada, famed for his ample appetite, went spear-fishing and returned with *isagi* fish to supplement dinner.

The final day of cycling had us continuing east along the coast. We stopped briefly at a Dairy Queen, across from the *pachinko* (pinball) parlor in the resort town of Kushimoto. There we bumped into our first foreigner in days, Dale, who was slopping down some soft ice cream. A Texan, she told us she was sailing around the world with her husband, on a 47′ trimaran called *Cherokee*. When did you start? someone asked. "Nineteen eighty-five," she twanged. "When will you get there?" I followed. "Never—we just plan to keep on sailing into the sunset. We're never going to stop. When we run out of money, we dock for a few weeks and take on local jobs, then we keep going. I hope I never see Texas again. And Japan, too. It's too expensive."

Not long after the Dairy Queen we stopped for a group photo at the coastal rock formation Hashi-kui-iwa (bridge pillars), a row of about thirty rock columns chiseled by the sea that looked like a procession of hooded medieval monks. While waiting for a tour bus of Japanese to load, I filled my waterbottle with tea, and watched as Tada stuffed his with sardines. Noticing my stares, he explained he was born in 1944, in the midst of the

Opposite, top. Rice paddies are ubiquitous, as much a part of the landscape in Japan as the mountains and clouds.
Bottom. In a crowded nation, nature asserts itself from a crowded bamboo grove along Route 168.

war, when there was no food. The lack of nutrition had affected his size, he said, and his brain. Anything he did that seemed strange or wrong, I should blame on the war.

An hour or so down the sinuous coast road I stopped at another vending machine, this selling hard liquor: whiskey, scotch, sake, and such. Again I wondered how the Japanese kept these privileges from being abused, but reminded myself that too much knowledge is painful; it is in small knowing that one finds happiness, in that and in nature, and I bought a Coke and continued.

Finally we turned back inland, for the concluding eight kilometer uphill stretch. Tada warned that this was the toughest climb of the trip, and recommended we consider riding the support vehicle. Addie, an engineer from Virginia, jumped at the chance. Pamela, who had her original bike back, less a couple of critical gears, also opted for the truck. That left ten for the final stretch.

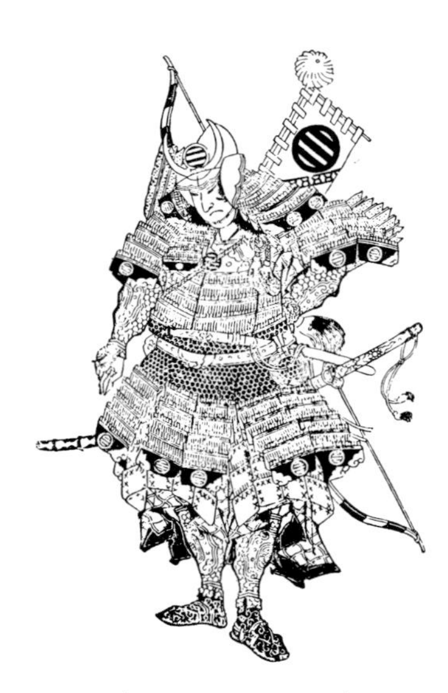

Japanese general.

We started out together, slipping through the backstreets of villages, riding along the edge of rice paddies, slowly working upwards. After a few kilometers I fell behind. Most everyone on this expedition was an experienced biker, except Peter, a member of the Central Park Track Club who evidenced no trouble with the rigors of cycling.

Soon I was alone, far behind the crowd. I continued my upwards quest, trying to match the sixty revolutions per minute Patrick had recommended, but rarely reaching fifty. I puffed and heaved, but kept on going. I began to wonder if I could make it, if the truck-not-taken was the wiser road. At one point I was certain I had pedaled more than ten kilometers, and wondered if Tada had fibbed to us about the distance. Some in the group were convinced he was a "cycle-pathic" liar, because few of the distances he had promised had been accurate. Then I saw the other bikes parked by a path at the seven-kilometer marker. I dismounted and followed a stairway of mossy stones down to a magnificent sight, Nachi no Taki, 436 feet in height, the highest waterfall in Japan. Japan's mythical first emperor, Jimmu, supposedly worshiped at these falls, and now I took out my camera to do the same. Since I had no tripod I moved an unoccupied bench to a vantage adjacent to a gravel-raked rock garden and set up my Minolta. After a few shots it became too dark to photograph, so I turned to leave. A monk appeared from nowhere, yelling like a New York taxi driver for me to return the bench. I bowed, asking forgiveness, returned the bench, and followed the ancient cedar-lined path back to my bike. The others had already left for the final kilometer to our ultimate destination. I had managed the last seven kilometers, so I figured I could handle the final fling.

But I quickly found it was something different. This last stretch was the steepest grade I had ever seen. It was steeper than San Francisco's Lombard Street, a switchback that curved upwards like the final pitch on Mount Everest. I clicked to the end of my gears and pedaled as hard as I could. My heart pounded. It was impossible to make a straight line, so I zigzagged back and forth, tacking the bike upwards. Several times I pivoted and cycled across the road perpendicular to my route, and then turned, like a good environmentalist, to recycle the same ground. It was a strain on my already tired body, and I thought of just quitting and walking the rest of the way, but I knew the others in front of me had made it. Besides, this was Japan, the estate of mind over body, of the unbending will, Zen and the art of bicycling. My lungs burned. I tried to think of other thoughts as I groaned around the next curve, tried to send my mind to another place. I thought of cheeseburgers, of my Australian blue shepherd, of forks and spoons, of CNN and hot buttered popcorn. And I pumped my legs, legs in great pain. Slowly I continued, erasing all thoughts of quitting. I thought of blue eyes, the Sunday paper, of large terry-cloth towels, of room service. I tried to imagine the sound of one hand clapping, of one foot pedaling. The sweat from my brow streamed into my eyes, blurring my vision. And suddenly I saw the sacred summit and our entire group gathered, awaiting my arrival. I straightened

Opposite, top. Sweeping the path to enlightenment at the Sonsho-in Temple in the wilderness of the gods, at the southern tip of Japan's main island, Honshu.
Bottom. At the Seiganto-ji temple visitors find both austere architecture and tacky trinkets.

見晴台

奉納
西國第壹番札所
那智山観世音菩
寄進
奉納
西國第壹番札所
那智山観世
寄進

the bike, and with a last burst, rode like the wind through the bright red *torii* gates into our cheering group, right up to a monk who was shooting my arrival with a video camera. "Did you make the final kilometer without getting off the bike?" asked Valerie. "Yes," I puffed, bent over to catch my breath. "Amazing." she continued. "You and Suzanne, the smallest person on the trip, are the only two who made it." I thought she was joking, that all the veterans had had no trouble making the final pitch. But she wasn't. Somehow I had bloodlessly ascended to this lofty perch, to the Sonsho-in Temple, a wilderness retreat overlooking a landscape of the gods.

We spent the following day wandering around the temple, bathing in its busy incongruities. Robed priests waved incense before a Buddhist cenotaph, and schoolgirls in middy blouses knelt and bowed deeply. A few yards away a forest of curio shops hawked *omiyage,* cheap trinkets: backscratchers, ear picks, bird whistles, pornographic key chains, and suggestive wooden carvings that a jaded tourist might bring home for his den wall.

But the ultimate schizophrenia came with our final banquet. The monks, with heads shaved much like the Yakuza's, decided to throw a barbecue at the base of the Seiganto-ji Temple, an ornate orange pagoda overlooking Nachi Falls and the various shrines of the compound. As two of the monks stirred the meat over the coals, another handed out ice-cold beers. Yet another recorded the events with his Sony camcorder while leaning against his Nissan Excel Skyline. As the night progressed the Karaoke machine was rolled out, and we were all enticed to get up and sing Beatles songs with the monks. But as I danced and crooned, my arms around the shoulder of a holy man on a mountaintop, there was a tranquility that flowed through me. The words to Paul McCartney's "Yesterday" became an elegiac ode, and I felt I was reaching the deep quiet of the Japanese spirit, getting closer to the flame of paradise, and feeling its chill. In my sozzled stupor thoughts fell away, or were gentled into something purer. And the air was filled with something more than music and wonder: peace, and the drift of meditation, among the chaos.

And somehow at that instant, like a profound dream that erases with morning, it all seemed to make sense.

Tasmania

Where Faust No Longer Paddles

It is more inspiring than a cathedral, more fragile and intricate than any painting or design, more involving than any playfield or theatre.
—Bob Brown, about the Franklin River

For eleven months of the year the Franklin is nothing but a brown ditch, leech-ridden and unattractive to the majority of people.
—Former Tasmanian Premier Robin Grey

On New Year's Eve 1980 I joined hands with a group of strangers on a bridge spanning the Stanislaus River in Northern California. It was a nonviolent protest of the closing of the 625-foot-high New Melones Dam just a few miles downstream. The steel floodgates had been shut and soon the Sierra's most magnificent whitewater river would be drowned. Over 100,000 people each year floated through the spectacular limestone gorges of the "Stan," splashing in its side creeks, exploring its Indian caves, running its exuberant rapids. I had been a part-time guide on the Stan since 1973, and I felt a temple was being ransacked. For ten years thousands of people fought this congressionally funded dam—through a state initiative, relentless lobbying in Sacramento and Washington, hundreds of demonstrations, and even near sacrifices, as some chained themselves to river rocks and bridge abutments as the reservoir began to rise. We were, finally, tilting at windmills. All the outcry, all the reason notwithstanding, the Army Corps of Engineers closed the floodgates in 1979, and by 1982 the most popular river run in the West was entombed under a 12,500-acre, twenty-four-mile-long "lake."

At around the same time, on the other side of the world, a similar battle was being waged to save a similar limestone-encased waterway: the Franklin River, in southwest Tasmania, an island shaped like a human heart floating beneath the mainland of Australia. Only Antarctica lies to the south. But this was a river scarcely known, a passage few had travelled. Four parties had attempted navigation by canoe or raft from 1951 to 1976, and fewer than five hundred had negotiated its rough waters by New Year's Eve 1980. And isolated Tasmania, traditionally governed by populist prodevelopment politicians, was suffering from high unemployment and a sagging economy. It was widely believed the cheap electricity a troika of dams could provide would attract much-needed industry and provide a new affluence. Yet, somehow, in a rare sleight of logic, the fight to stop the dams on the Franklin became a *cause célèbre* among Australians. Thousands rallied in peaceful protest. A government was toppled. And, miraculously, the river was saved, and is now protected as a UNESCO World Heritage Site. Tasmania's last wild river had become a locus of one of the most hopeful political changes of this century.

Now, ten years after my little protest on the Parrot's Ferry Bridge across the Stanislaus, I decided to visit the Franklin, and to see if I could meet Dr. Bob Brown, the man who had led the movement to save the little river in a faraway land.

Opposite. A waterfall at the misnamed Pig Trough waves its plume as it tumbles in two stages over sheer cliffs.

Two days after Christmas I met John, twenty-five, Andrea, twenty-five, and Molly, twelve, in the driveway of the Sheraton Hotel in Hobart, Tasmania, not far from where 20,000 had marched against the damming of the Franklin in February 1983, the largest land conservation rally in Australian history. John, slender and vulpine, was the leader of our little expedition, and had made twenty-five trips down the Franklin; Andrea, though a world champion kayaker, was the relative newcomer with ten trips. And Molly was the rusted, dented 1978 Mazda van they used to get around town.

Two days later, with a party of twelve, we travelled the Lyell Highway to the put-in of our trip: a tributary of the Franklin, the Collingwood River. It was a blustery, wet, cool day. The gage on the concrete bridge abutment read 1.2 meters, giving me pause. The guidebook I was carrying stated the river shouldn't be attempted over one meter. John, unconcerned over the water level, merrily rolled out the inflatable rafts, two twelve-foot-long Avon Adventurers, to be piloted by John and Andrea, and a dinky ten-foot-long self-bailing black SOTAR (State of the Art Raft), named *Darth Vader,* which I would paddle along with the other two Americans on the trip, Steve Marks, a Hollywood talent agent, and Pamela Roberson, the trip photographer. Then John gave us his orientation, and we all buckled into our one-size-fits-nobody plastic helmets and slipped into our wet suits and booties.

All except John, who was seemingly impervious to the weather. He wore only a pair of shorts over his striped long thermal underwear, and wool socks in his sandals.

By 12:30 I was sailing downstream in a frail chip of a raft on a river of primal intensity. Above, a black cockatoo, a harbinger of bad weather, screeched in the wind. Below slurped a river that appeared as dark as the bird, the result of tannic acid eluted

Below. The Franklin runs dark as rum, the result of tannic acid eluted from the nearby buttongrass plains.

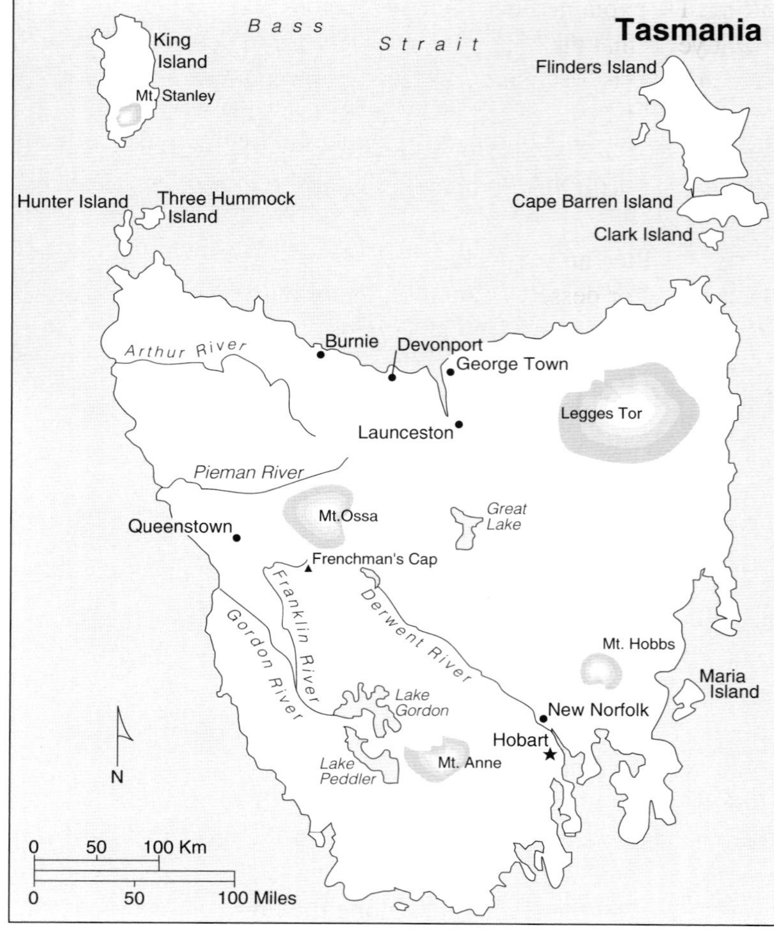

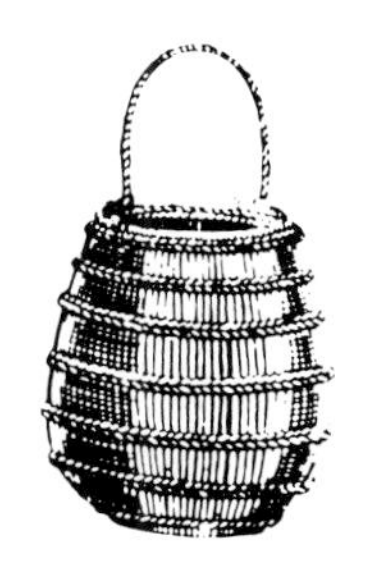

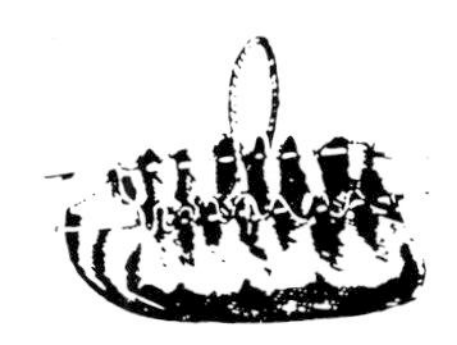

from the nearby buttongrass plains. The solemn color of the river matched my mood. Franklin, my dear, I was damned scared of what lay ahead.

After a few hours bouncing through modest rapids we came to the confluence with the Franklin proper. At this juncture the Franklin had scrawled 3,000 feet down from the ice-rimmed tarns of the Cheyne Range in Tasmania's central highlands. It ran the color of old beer, not especially appetizing, yet I dipped my cup knowing it was among the cleanest water on earth, flowing through a region with the purest recorded air on the surface of the planet. Here we found a party of Australians with two cheap yellow Taiwan-made rubber-duckie boats and a deflated gray dinghy called a Sea Eagle sprawled on the gravel bar. The Sea Eagle had punctured on the only rapid of consequence on this first stretch, Sticks and Bones, and the castaways had been trying to repair it for the last day and a half. They asked us for glue. John said we had too little to spare, and suggested they consider hiking out, as a trail up the Franklin was the last easy exit for the next seventy-five miles. They promised to consider the option, and we pushed off down into the bosom of the mother Franklin.

We continued through several more sets of boisterous, white-laced rapids with evocative names like Gordian Gate and Boulder Brace. Finally, we pulled over for camp on the eastern bank at Angel Rain Cavern, a grotto festooned with quivering ferneries and a perpetual curtain of misty rain.

Ours was a large party for the Franklin. Most groups are half our size, because of the shortage of campsites on the steep, dense banks. We had to squeeze under the overhangs, a prospect that, because of my legendary snoring, pleased none. Though driftwood was in abundance, John assembled two gas stoves and began to prepare dinner. I asked him why he didn't make a fire, something that could help warm wet souls after a day of rafting. He explained that so many serious fires had devastated the Franklin catchment in recent years that the Wilderness Society had requested rafters to stick to stoves. Privately, Andrea, who is Tasmanian and has been running the rivers of her island since seventeen, told me she thought John, who hailed from the Australian mainland, was perhaps being overprotective. She felt wood fires, if properly attended, had their place, especially on a river littered with logs, branches, and forest debris, obstacles that presented a major danger to boaters.

After an early dinner of deep sea trevalla in a lemon and butter sauce, and cheesecake for dessert, John took us for a nature walk. He pointed out sassafras, the fragrant white flowers of the leatherwood, kerosene bush, the glossy leaves of native laurel, Christmas bushes, and scaly-bark tea trees. I slipped into my bag soon after dark, and listened for a while to the regular *plink* of droplets, the Angel Rain; to the faint call of the tiny woodland bats; and then to someone snoring at jet-engine-decibel level. I smiled and easily slid into sleep.

The next morning was gorgeous—a narrow slice of azure sky above us, and below a fairy-tale river whose level had dropped, improving my mood. On the water I watched as migratory martins darted from their cliff-side nests, and grey fantails flitted, snatching insects, across the surface. Not far downstream we met our first impasse, a grand old log, anchored on boulders and blocking the river from bank to bank. It could have been there a century, maybe more. We wrestled the boats over the fallen log and continued down a corridor of vast, scrub-hung cliffs. Soon thereafter, the river veered south past the Raglan Range, then picked up the Loddon River. There we saw our first Huon pines, endemic thousand-year-old trees, leaning from the banks. In fact, the first river runners were loggers who came up the river looking for these close-grained trees, and then floated them downstream to be hewn for ships and coffins. From 1900 to 1950 over 90% of all the Huon pines along the Franklin and its tributaries were felled.

Not far beyond we careened into the Overture, a rapid that spilled into the shadow-throwing cliffs of Descension Gorge, and that in turn spun us into the deep, peaceful quartzite chasm called the Irenabyss (named by the aforementioned Dr. Brown, and meaning "Chasm of Peace" in Greek). Irenabyss . . . I loved the name, and rolled it on my tongue like taffy as we crept through the smooth, water-polished walls of the canyon, 150 yards long and 100 yards deep, and narrowing to barely a boat's length in one spot. It was a crooked course that finally blossomed to a quiet sunlit pool at the mouth of Tahune Creek, where we pulled in and made camp.

Opposite, top. The author piloting the *Darth Vader* down the rapids of the Franklin. *Bottom.* Camp under the overhang alongside Newland's Cascade, the longest rapid on the Franklin.

After hanging my clothes out to dry on the wind-pruned branches of a pine tree I crawled out on a rock shelf that overlooked the Irenabyss. It was to have been, John said, the site of a proposed dam on the Franklin. It didn't seem possible that such a beautiful place could have been marked for death, that the whispers of the river could be replaced by the hum of turbines. As I looked into its depths I sat hypnotized by long lines of foam that curled like cream across the burgundy surface of the river. An oblique ripple broke my reverie, a mountain trout catching an insect, and then I heard an unfamiliar voice.

Looking up I saw a woman in a drab green Ranger suit with a platypus patch. She introduced herself as Rima Truchanas, a seasonal track ranger. She had just hiked in from Frenchman's Cap, a giant exposed quartzite pyramid that loomed up the valley. As it turned out Rima was the daughter of Tasmanian legend Olegas Truchanas, a Lithuanian-born bushwalker. He had led the failed attempt to save the upper Gordon River from a dam that destroyed a jewellike natural lake called Pedder. In 1972 he had written: "If we can accept a role of steward and depart from the role of conqueror; if we can accept the view that man and nature are inseparable parts of the unified whole—then Tasmania can be a shining beacon in a dull, uniform and largely artificial world." Shortly after he wrote those words Olegas Truchanas drowned while canoeing the river he couldn't save, the Gordon. Rima was ten. Now his daughter carried the torch, working to keep the wilderness pristine, citing hikers for not practicing "minimum impact bushwalking," and spending days, sometimes weeks, alone in the sanctuary of a saved river canyon, the Franklin, a tributary of the Gordon.

Tasmanian devil. Tasmania owes its name to Abel Tasman, the Dutch navigator who sighted the island in 1642. Until 1856, it was known as Van Diemen's land, after a later explorer.

The next morning we took off on foot to climb the massive block of Frenchman's Cap, called the most majestic mountain of Terra Australis. It was a slog. The first hour was through thick, steep, twisted, tangled, leech-infested, bracken forest. Then it got tough. We waded through spiky buttongrass, knife-edged scoparia, over irregular, sharp schist, and through stunted alpine heath with no shade from the hot sun. After two hours we stopped and looked back down the valley, back down over the eucalyptus, the great dark myrtles, and celery-top pines that rimmed the peat-brown river canyon. Ah, wilderness! We couldn't decide if we should celebrate or curse.

Several interminable hours into the tramp we stopped for a rest beneath a wall etched by eons of weather. Days here in the Roaring Forties (we were 43 degrees south latitude) were usually overcast and cool from the low-pressure systems of Antarctic origins. But today was different, hot and sunny, not a cloud in the wide vault of sky. Warned only of the typical weather, none of us had thought to bring water bottles, and I, for one, was dying of thirst. Scanning the wall I noticed a trickle emerging from a patchwork of moist moss and lichen, a couple of drops every few seconds, and I tried to intercept a drink with my mouth, but couldn't press my face close enough to the rock. I tried to catch the water with my hand, but the flow was so scant the water seemed to evaporate before I could drink. Then I thought to empty a film canister, and for long minutes I held it against the wall as it filled, and then I drank the precious sweet water.

Sometime in midafternoon we reached the plateau. For a couple of hours I stumbled over bare patches of scree and landslide debris under the full bore of the southern

Sketches done by Tasmanians on a tree, representing (from top) people and animals, sun, moon, snakes, and five people in a boat.

sun, searching for cairns to mark the way. Several times I found myself unsure of the route, and circled the area looking for a pile of rocks as a sign, but could find none. Then I saw John far ahead knocking over cairns as he walked along an arete. I caught up with John and asked why he was destroying the trail markers. He explained that Rima Truchanas had convinced him cairns were damaging the delicate alpine ecosystem, as they encouraged hikers to all walk the same route. Too many boots created a gully, which in turn unnaturally drained the landscape, beginning a process that would irrevocably alter the face of the mountain. Rima felt "minimum impact bushwalking" meant avoiding trails, walking in random routes along the plateau so erosion would be more evenly distributed. That was fine, I thought, until somebody got lost and walked off the mountain, but I was too tired to argue. Instead I made sure I stayed close behind John as he knocked his way to the North Col, the saddle separating Lakes Tahune and Gwendolen, at the base of the summit pitch to the Cap. From the Lyell Highway the blanched 350-yard-high cliff-face looks like a hat worn by a French cook. From my vantage it looked like Yosemite's Half Dome, radiant with quartzite whiteness, and quite imposing. Beneath this gusty aerie I could see our intended campsite alongside Lake Tahune, and just beyond a sweep of hundreds of acres of ghostly white trunks, the skeletons of King Billy pines, ravaged by a fire in November 1966. While admiring the view from the North Col, Pamela lost her grip, and her Lowe-pro camera bag bundled down the steep slope towards oblivion. But John, with lightning reflexes, bounded after the avalanching bag, and a hundred yards down the slope, at a brink where the earth abruptly snapped off, he caught the bag. My estimation of John's prowess ratcheted up several notches. Pam couldn't thank John enough, and called him her rescuer down under.

Four of us, John, Steve, Adam (a paddler on Andrea's raft) and me, decided to attack the summit that afternoon while the rest descended to Lake Tahune to set up camp. We followed a trail that switched around Precambrian sentinels and small outcrops, passing a drift of snow that had a group of other hikers sledding down on their bums, and up the scarp of the bald bluff. Finally we could climb no higher. We were on the roof of Tasmania, 3,958 feet above the sea. A chill wind buffeted us, but we hardly noticed, wrapped in the blanket of one of the grandest panoramas in the Pacific. Each direction tapped at our shoulders. To the east Mt. Ossa (Tasmania's highest peak at 4,436 feet) and Barn Bluff shone like distant planets. Macquarie Harbor, where the Franklin delivered its waters, lay to the west. And farther still we could see the glint of the Southern Ocean. Dark cirque lakes cupped beneath sawback mountains surrounded us. The only sign of man's handiwork was the barren cut of an HEC (Hydro-Electric Commission) dirt road winding its way down Mount McCall to the second Franklin dam site. It was beauty betrayed, a scar slashed across Nature's cheek.

We made it down to the dark waters of Lake Tahune just before sunset. The map showed we had walked only two and a half miles, but it had taken us eleven hours to reach this spot, and we were exhausted. But it was New Year's Eve, and we refused to be party poopers. From the heavy packs we had strained to carry to this loft came colored balloons and party streamers, which we slung between the trees. Pam produced noisemakers and equally dignified party hats. John and Andrea whipped up a Mexican meal, topped with sweet Paulovas (sweet pastries) for dessert, and even nature cooperated: the full moon rose, and a heavy fog dipped into the basin, hanging on the pagodalike branches of the pines. But despite our intentions, we were fatigued, and it soon became apparent we wouldn't make it to the stroke of midnight. We opted instead to celebrate the Fijian New Year. Fiji, just our side of the International Dateline and some 2,500 miles northeast, would ring in the New Year at 10:00 P.M. our time. We donned our paper hats, John and Andrea changed to Fijian sarongs they had brought up the hill with them, and at the stroke

Opposite. The waterfall just above Rock Island Bend splashes into a paradise of soft water ferns and velvety moss.

Opposite, top. Limestone cliffs worn in fluted curves line the lower Franklin. *Bottom left.* The Franklin River water is reportedly the cleanest on earth, flowing through a region with the purest recorded air quality. *Bottom right.* Most Franklin rafting parties are small, less than a dozen, because of a shortage of campsites on the steep, dense banks.

of ten we all cheered and shared a couple of bottles of Seppelt Great Western Champagne. It was a heady, happy moment, but my thoughts wandered. It had been ten years since my New Year's Eve vigil on the Stanislaus, and that night stayed in my mind. I silently toasted the submerged sister river far, far to the northeast.

"No worries, mate," were the first words I heard in 1991. It was John, outside my tent, giving his signature answer to almost any question. The question turned out to be, "What do we do without Andrea?" as with dawn she was nowhere to be seen. We soon found out. Andrea, with almost super-human energy, had decided long after the rest of us had collapsed into sleep to climb to the top of Frenchman's Cap and bivouac under the full moon. She was there still, but joined us later, her blue eyes sparkling, at lunch on the saddle. From there we all tumbled down the slopes to the Franklin.

Back at the Franklin camp we came upon the rubber duckies we'd met at the junction two days earlier. The Australians hadn't taken John's advice, and instead had limped downstream with a jerry-rigged Sea Eagle. Once again they begged us for glue. This time John, knowing the worst was ahead, shared the precious epoxy, and wished them luck.

The next morning I watched as the river was reborn from its soft cocoon of early morning mist. The rhythms of the water seemed to beckon. Flowing west and then south, the Franklin continued to twist between the serried ranks of the Engineer Range and the foothills of Frenchman's Cap. Downstream the river funnelled between a bouldered interstice at a place called Toothpaste Tube. We wedged sideways and came close to capsizing, but somehow managed to squeeze through. But not without consequence. We were to camp that night at The Crankle, the middle of a mile-long loop during which the river flows in every direction of the compass. But as we paddled to The Crankle the floor of our little black boat lost its air, and the ice-cold water in the bilge felt like a vise on my feet. We quickly beached and turned the raft over. The floor, perhaps from squeezing too hard through the Toothpaste Tube, had started to delaminate. The vulcanized seams were coming apart, their repair too difficult to attempt on the river. So Steve, Pam, and I were resigned to a bottomless and hence much slower craft for the rest of the trip, and a bilge of chilly water for our feet.

We made camp high up the bank, as the Franklin is a river that can rise as much as thirty feet in a few hours, a sight John had seen once with his own eyes. I pitched my Macpac dome tent just beyond an elaborately carved dead Huon pine, a tree of incomprehensible age decorated by a salvage logger. Louise, a Londoner now working for ABC (Australian Broadcasting Commission) in Melbourne, went to the loo and came face-to-face with a brush wallaby, but when I ventured, it had disappeared into a tangle of dogwood. Then, during dinner, a weasellike carnivorous marsupial called a quoll wandered into camp. He poked his way through our bags, into the wash bucket, and even climbed on one of the rafts, as we watched in amusement. It was clear this was a camp enjoyed by wildlife, and someone wondered aloud if perhaps the Tasmanian Tiger, last seen in 1938 not far from here, might not be somewhere behind us, watching us as we ate.

That night, as I left my tent to answer nature's call, I swung my flashlight into the woods and saw a dozen amber eyes looking back. I froze for a second—perhaps a pride of Tasmanian Tigers? Then I swept the light over the camp and saw there weren't a dozen eyes, there were scores. These were innocent glowworms staring me down, not quite ready to pounce.

The day dawned misty. The bent trees across from camp seemed to inhale fog from the wind. A fish-seeking cormorant made a low, silent sweep above the river. It was a day packed with rapids and tributary waterfalls; at every turn the river gained in volume and authority. At lunch (which, as always, consisted of linseed bread, mango chutney,

vegemite, scrogin (trail mix), sesame tahini, and, thank God, peanut butter), John took me into the forest to have a "squiz" (a look) at a small plaque attached to a tree: CANOEISTS HENRY CROCKER, JOHN DEAN, JOHN HAWKINS, TREVOR NEWLAND 31-12-58. Hawkins and Dean had made two unsuccessful attempts at running the river, in 1951 and 1957, and both times, with canoes destroyed, they hiked out less than a third of the way down. In 1958 they finally made it, using 13½-foot fiberglass canoes, and took the honors as the first to successfully navigate the Franklin top to bottom, almost thirty-three years to the day before our little sortie.

That afternoon the river narrowed and gathered momentum, and we met the dreaded six-mile-long Great Ravine—the largest gorge in Tasmania, and the bane of Franklin rafters. The Franklin trip is a lottery. If the water is high, and weather bad, rafters can spend days, even weeks, in this section, making "high portages" in which all the gear, including the boats, must be lugged hundreds of yards up steep cliffs and negotiated through the raw terrain. Luckily the river was low, the skies still clear, so we were able to tackle the relatively easy "low portage" route with the first of the four infamous portages. It took two hours to carry our personal kits and the seven watertight plastic brewing barrels containing our food and common gear around the chaos of rock and water called The Churn. John and Andrea, meanwhile, pulled the empty rafts up a cliff face, then dropped them into a boiling knot of water below two unnavigable waterfalls. John especially was uncannily agile on the slippery rocks, leaping like a wallaby from rock to raft without the slightest hesitation. I felt like an old man trying to follow John, slipping and sliding on the bare, scoured surfaces he flew across, using both hands, knees, and feet to find purchase where he only needed a toe. I was once a river guide, and thought of myself as fleet-footed and agile, but nothing like John, who seemed more fly than human. The last part of the Chute was a Class V rapid—extremely difficult, long and violent rapids—and we decided to attempt it. After fastening the gear back on board, we piled in the rafts, took positions, and shot this stretch fast as a bullet through a gun barrel.

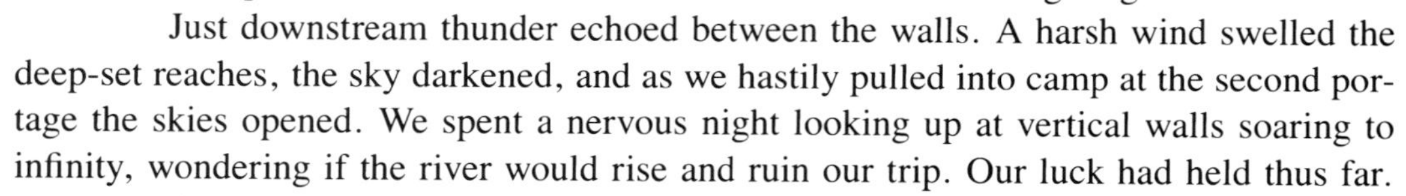

Just downstream thunder echoed between the walls. A harsh wind swelled the deep-set reaches, the sky darkened, and as we hastily pulled into camp at the second portage the skies opened. We spent a nervous night looking up at vertical walls soaring to infinity, wondering if the river would rise and ruin our trip. Our luck had held thus far.

Tasmanian woman.

Against all odds, Friday was another pretty day. But John had attached his good-luck charm, a plastic Ninja Turtle doll, to his helmet, so we knew there was a world-class obstacle course ahead. The second portage, three hundred yards around the serpentine Coruscades, was tough but not overwhelming. Again John and Andrea handled the boats. At the bottom, in the half-light of the ravine, we filtered through the Faucet, where Adam washed out. He looked shaken when we finally pulled him back aboard, and it reminded me that risk was a constant companion on this trip.

On Christmas Eve, 1840, surveyor James Calder was in this area, cutting a track for Governor John Franklin's expedition to cross the southwest of Tasmania. Calder examined this "abrupt country" for two days:

> I tried to lead the road across at several points, but was thwarted by the intervention of a tremendous ravine. . . . This locality presents no other view but that of a sterile wilderness, and scenes of frightful desolation. The great ravine is far too steep for travelling, and not to be crossed without excessive fatigue and risk. . . . I twice got to the bottom of this hideous defile, but was at last forced to relinquish the idea of a direct course . . . utterly disgusted with the adventure. A large and furious torrent flows through it, which, collecting all the water that falls on a wide extent of mountainous country, emerges from the glen a large and beautiful river. I called it the Franklin.

Thunderush was the big portage, one that had us negotiating the right bank up rope ladders, down steep gullies, across catwalks, around spurs, and along cliffs for a tortuous mile, before dropping into the river just one hundred yards from the start of the raucous rapid. John, Andrea, Adam, and another of Andrea's passengers, Bill, ran the last section, where the whole river spouted like siphoned soda over a worn rock, and then spat into a bottleneck called the Wedge of Eternity. All three runs found the boats stuck in the Wedge, where they were manhandled out and delivered into the Sanctum, the calm water at rapid's end.

The last Tasmanian aborigine died in 1869.

The final portage in the Great Ravine was around an aptly named piece of whitewater, The Cauldron, a boiling brew that had claimed the life of a guide in 1984. Herein lies the paradox of wild rivers. They are timeless and enormously strong, totally dominating human visitors, capable of killing them; yet they are vulnerable, subject to the whims of people in faraway cities, exposed to injury by chainsaws, mutilation by bulldozers, and death by concrete and landfill.

John once again masterfully set up the complicated logistics of the portage, with strategically positioned throw lines and safety crew. In this particular exercise the rafts were first dragged down the rocks on the left shore, then paddled across an interlude of calm water just above the most lethal falls. The final act was a pour over a steep aerated pitch where John rode the boats like a cowboy taming a wild horse. With the successful delivery of each raft John would let out a war whoop, as his heroes Arnold Schwarzenegger and Dave Foreman (founder of the radical environmental group Earth First!) might have done, and then he shook his fist in victory.

Once delivered through the Great Ravine, the pressure was off. John, a vegetarian, spent hours at the Rafter's Basin camp preparing one of his favorite meals, a pumpkin curry, with plum cake pudding for dessert. The sunset that touched our tableau that night was magical, almost as colorful as the ground parrot that flitted through camp. I was spellbound by the unvarnished beauty of the Franklin.

Just below Rafter's Basin we sighted a shapeless wad of yellow draped over a bronzed rock. We paddled closer. It was a rubber duckie, like the ones we'd see at the confluence. But it couldn't be one of those; we were certain, well almost, that they were still behind us. Soon after we passed the proposed site of the second impoundment in the grand 500-million-dollar hydroelectric scheme for the Franklin. Up the western bank we saw the lacerations of bulldozers, where not long ago the silence of the forest had been splintered by the crash of falling trees. But now the silence had returned. The Franklin to this point had been unsullied by any hint of man, no fences, farms or houses, no membrane of human cultures stretched over its terrain. It remained a wilderness indivisible, a singular place where natural cycles persisted undisturbed. The bulldozer scratch was a sobering reminder of how close a shave the region had undergone. If the 600-foot-high dam had been built, the entire Great Ravine would have been flooded.

The day was supposed to be one of relative calm, but perhaps because we knew we were beyond the worst of the river, we weren't as cautious as we should have been. In one unnamed drop I lazily steered the raft over a submerged rock, rather than put the muscle into paddling around, and I was tossed forward against a camera case. When we emerged into the languid pool below I saw water in the bilge swirling red, and I looked to my left leg, where a foot-long gash oozed blood. It was a shallow cut, but an unnecessary one. Still, I didn't pay any more attention when the next rapid hit, and this time I steered *Darth Vader* sideways onto a rock midrapid, and the boat began to wrap, the worst scenario for an inflatable raft. The downstream side of the boat began to ride upwards, and the bow and stern bent around the rock. Upstream water began to pound inside, plastering the whole affair with tons of pressure. Steve and I scrambled onto the precarious perch of the rock,

where we yanked and pushed and kicked, but we couldn't release the raft from the river's tight grip. The boat continued to slowly rise up the rock and flatten itself. A few more inches and the boat would be lost, and we would be stranded on a slippery roost in the middle of the torrent. So, with a last Herculean effort we pulled the stern downstream, and the raft inched along with us. Again with all our effort we pulled, and suddenly it was set free. I jumped aboard. It rocketed downriver, leaving Steve on the rock for a long second. He leapt for the raft, reaching the bow with one hand, where Pam pulled him in. We had felt the power of the beast, and I vowed no more complacency.

After lunch John led us into the woods, where he showed us a magnificent felled Huon pine, the largest he'd seen on the river. Its gnarled trunk was obviously very old, perhaps a sapling when Christ was born. Yet for all its years it probably had seen very little change along this special corridor, until the day a logger took an axe to its trunk.

Maneuvering from shore with long ropes attached to the nose and tail of the rafts, we moved them through Pig's Trough, where a rafter had drowned in 1981 in a pitch of aerated water. Just beyond, the river wound around the defiant and lovely Rock Island Bend; a color photograph of the Bend had run in several major Australian papers and is credited for helping to turn public favor towards saving the river. The hotel-sized rock in the bend was trimmed with moss and myrtle and piled high with smashed timber, the size of which gave some idea of the forces that had thrust them there. We climbed up to see the Pig's Trough waterfall, which splashed into a paradise of soft water ferns, velvety moss, and palmlike Pandani plants. I climbed to its base and played Adam in Eden, just as early convict loggers more than a century ago would bathe here after a dirty stint in the woods. Then we ran the slalomlike Newland's Cascade, at a quarter-mile from tip to trough the longest runnable rapid on the Franklin. I botched the entry and caromed down the wrong side, through improbable chutes and over unlikely staircases, but somehow emerged right side up.

The highlight of the two-night layover at Newland's Cascade came the next day when we looked up to see the rubber duckies bouncing through the rapid. They had finally abandoned the leaky Sea Eagle raft, and were now packed into their two remaining bathtub-toy-type rafts. They didn't even scout the rapid. Just crashed and wormed through, folding over rocks, punching through waves. It was testimony to how forgiving this river could be.

The next day, the coldest yet, we continued our river trek, dropping through the occasional shallow, shingly rapids. John, oblivious to the Gothic weather, continued wearing his uniform of shorts over long underwear and socks and sandals, while the rest of us, layered in wet suits and outerwear, shivered in the drizzle. The terrain had changed now. The valley had widened and flattened. Low leafy banks and occasional blue-grey limestone cliffs, worn in fluted curves, lined the broad water. We dawdled through tranquil, deep, dark-brown pools, perfect mirrors, glinting copper-colored where the sun penetrated. The raft in front of us sighted a large platypus, and we all watched as a set of white cockatoos flew across the sullen grey sky.

That night John broke his rule and built a fire. He confessed that he had mistakenly filled one of the carefully rationed gas canisters with dish soap, and so we were out of fuel. The real fire would give us a chance to dry clothes and warm hands and feet over a flame. John boiled the "billy" (pot) full of tea, or so I thought as I fetched a cup and drank. I spat it out—it had no taste. He hadn't put the tea bags in just yet, and the color of the Franklin water is so much like tea it was impossible to tell. So, I settled for Milo (an Ovaltine-like drink) and warmed my hands around the mug.

Clouds gathered ominously the final morning. Over coffee I paused at a passage in a book I had carried down the river, a collection of local environmental essays called

Opposite, top. Not far from here a dam was planned, one that would have drowned this fairy-tale section of the river under a stagnant reservoir. *Bottom.* The banks of the Franklin are a moist, misty riot of ferns and temperate rain forest.

Overleaf. Governor Franklin, on his 1842 expedition, described his encounter with the river: "It was a scene of heaven directed gratitude—of joyful exultation."

Below. The Irenabyss Gorge narrows to barely a boat's length. It means "Chasm of Peace" in Greek.
Opposite. Portaging through The Great Ravine, the largest gorge in Tasmania, six miles long and churning with four major unnavigable rapids.

The Rest of the World is Watching. It quoted Goethe's *Faust:* "Grey is all theory; ever green is the tree of life." So inspired, I packed away my usual river gear and instead wore shorts over my long underwear, and my Teva sandals. An hour downstream, on this ribbon that seemed to run back through time, we met the feisty currents of the Jane River, named for Lady Franklin, who gamely joined her husband on the eponymous Franklin Expedition of 1842, during which she was carried by convict porters on a palanquin. A few turns beyond we scrambled up the soft bank to explore the Aboriginal Kutikina Cave, a natural museum archaeologists believe housed the southernmost habitation of people anywhere on earth during the last Great Ice Age. Behind a proscenium of ferns were two cryptlike rooms lined with pimply stalactites and dark flowstones. A barrier kept us from disturbing the stone tools and wombat bones, among the world's oldest leftovers. It was daunting to know that Tasmanian Aboriginals, some of whom had dined in this den 20,000 years ago, were hunted like vermin by the European settlers, and that the last full-blooded Tasmanian Aboriginal was "exterminated" in 1876, less than seventy-five years after the arrival of Western "civilization." Perhaps we had yet to civilize. At the height of the Franklin controversy, some of the pro-dammers had threatened to blow up Kutikina Cave, and "burn down the Southwest."

After lunch we were again reminded that we were still in a wilderness, where a man, at his mischance, could perish. We lined the rafts through the last bedrock rapid of the trip, Big Fall, a dangerously deceptive six-foot drop with a powerful barrel-shaped hydraulic at its base that sucks wayward rafts and people into a recirculating hole. To date two have drowned on this last lick of the Franklin. From here on down the river is navigable by motor boat, though that's been forbidden for the past couple of years because of the shore damage from wakes. During the 1842 expedition Governor Franklin travelled up the Franklin to this point. His diarist, Mr. Burns, recorded the setting: "It was a scene of heaven directed gratitude—of joyful exultation. Here the glorious masses of light and shade so peculiarly beautiful in all Tasmanian landscapes were seen in their fullest and

finest effect." Though the light was undeniably beautiful, at that moment our gratitude was haven-directed—the haven of our take-out, where we could change to dry, warm clothes and sip a palliative brandy.

Finally, late that afternoon, we paddled past Pyramid Island, and joined up with the Gordon, Tasmania's Old Man River, its largest and longest waterway. Bob Brown once camped on this tiny isle. Likely fleeing convicts did as well, perhaps James Goodwin and Thomas Connelly, two desperate men who paddled upriver in a Huon pine canoe on a summer's night in 1828 to escape the horrors of Macquarie Harbor Prison (the *ultima thule* of the penal system). Of the hundreds who tried to escape, Goodwin and Connelly were among the five known to have made it alive through this wilderness. Not far up the Gordon was the Scotts Peak Dam, which drowned the beautiful Lake Pedder in 1972. UNESCO described its destruction as the greatest ecological tragedy since European settlement in Tasmania.

It was a glum and hunched paddle past the statuesque blackwoods, the lofty gum trees, and the riverbank acacias towards our conclusion. Overhead a sea eagle watchfully soared. For the first time our backs and shoulders felt the strain, and in my uniform of long underwear and shorts I was cold, and cursing Goethe. Two hours later we pulled into Sir John Falls, our take-out, some thirty miles east of Macquarie Harbor, where the waters that began as the Franklin debouch into the sea on the southwest coast. Just downstream was the blockade site against the construction of the Gordon-Below-Franklin Dam, the place that made the Franklin and Bob Brown notorious. If history had played a different hand our landing would have been 250 feet under water.

My last day on this small island at the end of the earth I said goodbye to John and Andrea. As if to prove the Franklin wasn't tough enough, John announced he was leaving in a few days to spend the winter in Siberia. Andrea told us that she had accepted a teaching job, and would not be able to raft any more that season.

Then I went to see Bob Brown, the fountainhead in the movement to save the Franklin. I wandered over to the Parliament building, and through a labyrinth of corridors until I found the forty-six-year-old eco-politician in a crowded room filled with environmental paraphernalia. I had heard so much priestly praise of Bob Brown—he had won the Goldman Environmental Award in San Francisco just a few months earlier—that I approached him with some trepidation. It was defeated, however, in less than an hour. His humanity and quiet sincerity, his transparent love for the wilderness, could not be counterfeited.

Two weeks after being arrested at Sir John Falls for his nonviolent direct action to save the Franklin, Brown had been voted Australian of the Year. Less than a week later, on January 5, 1983, he was elected to state Parliament as an Independent Green. Now he and four others of the Green Creed hold the balance of power in the government, the only

Opposite, top. Sir John Falls, the take-out of the Franklin River expedition.
Bottom. Camp at Lake Tahune, on Frenchman's Cap, on New Year's Eve.
Below. The float plane that delivers rafters back to Hobart, the island's capital.

A Tasmanian boat used for travelling short distances only. The Tasmanians had a great fear of water, so their boats needed to be suitable for no more than crossing rivers or small stretches of water.

environmental party to do so in the world. Jug-eared, long-faced, tall and thin, his body all sharp angles that didn't seem to fit comfortably together, Brown seemed as simple as the son of a country policeman that he is. He told me he had rafted "his old friend" the Franklin six times, the first in February 1976 in a gaudy rubber duckie. He hoped to go as many times again, but for now he was too caught up in other battles, such as a civil disobedience action he had scheduled the next week to protest the cable logging of the temperate rain forest along the Arthur River in the north of Tasmania.

I asked him why the Franklin was saved when so many other great wild rivers of the world had been lost to dams. He gave me a youthful, country-boy grin.

"I think it was a test of conscience in Australia. For two hundred years it had been 'open go' on the Australian environment by European settlers who had no recognition that it was a finite thing. There was always more over the horizon. And, there was a prevailing attitude that the most scenic and spectacular places were elsewhere, in Europe and North America. When we showed the Franklin through words and pictures it awakened people to the fact that we have this extraordinary beauty in our own country, and that this ancient country has an integrity of its own that should not be destroyed to meet transient human needs."

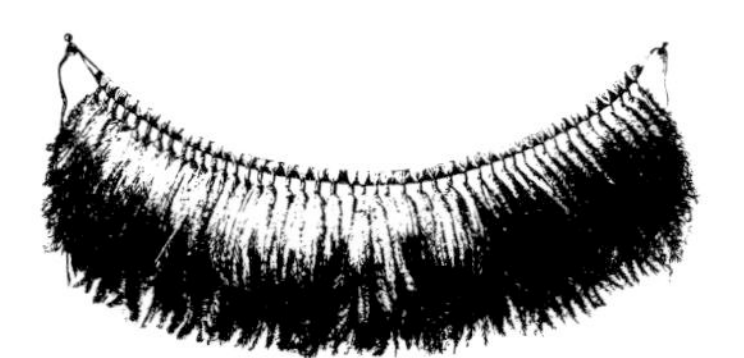

Dance adornment of Aborigine women.

Was the Franklin issue an environmental spin-off of something much deeper?

"Yes—I think it is a cameo of a growing world concern that we are on a small green planet, floating through space, and we don't quite know where we are going. But we do know that the planet is being rapidly changed by us and we are totally responsible for it. Unless we take control of it, we won't have a future, because we are faced with a real crisis of technological overkill."

Bob Brown's speech was as plain as his name. Tasmania is the whole world writ small, and if reassurance, meaning, and friendship with the planet can be found here, if a man like Bob Brown can rally the collective spirit and correct the juggernaut of "progress," then there truly is hope. The river had given Bob Brown almost transcendental strength, as wild places can. I felt that twelve others who had just rafted the Franklin were stronger as well.

When I returned to San Francisco I picked up the paper and found an article about the Stanislaus River. After five consecutive years of drought the reservoir that had once drowned the nine-mile-long upper canyon was so low that the river was resurrected, flowing with a pulse and filling the lungs of the canyon with fresh air. Rafters had been running the reemerged rapids, new grass and wildflowers painted the shoreline, and a pair of ospreys had nested near the water. But even more importantly, many of the people who originally supported the dam had now publically reversed their sentiments, had turned the telescope towards irredentism. There was a call for reexamination, and there was a chance that the dam might come down.

One day not long ago it rained, and while walking to town I stopped to watch the subtle battle between the water and rocks in the flooded creek near the road. There was no contest, I thought. Despite appearances, I knew that water would always beat stone.

the Trobriands

The Islands of Love

Chastity is an unknown virtue among these natives.
—Bronislaw Malinowski, 1922

Sex here is like drinking water.
—Bill Rudd, 1990

Like passing through a pool of light, it is a traveller's thrill to delve and then move on.

The ball was set in motion the day I realized I was closer in age to Ed Sullivan than to the Beatles when they appeared on his show. It was the eve of my fortieth birthday, and I noticed a certain grimness growing about my mouth. It wasn't winter that threatened my soul, but a long, dank autumn. I didn't want the typical party with limp balloons on the wall and all manner of condolences over deteriorating parts. I wanted to be out of town in open spaces, someplace fabled and exotic, someplace I had never been, only dreamed about. And as I spun my mental globe it stopped in the South Seas, off the eastern coast of New Guinea, on a fertile four-island archipelago set in a figlike pattern: the Trobriand Islands, often called "the Islands of Love." It seemed the perfect spot, especially when I heard that the annual Yam Festival, during which the locals shed all inhibitions to celebrate their all-important harvest, was held in late August, coinciding with my birthday.

Named for Denis de Trobriand, first lieutenant in France's 1793 D'Entrecasteaux South Sea expedition, the Trobriands were made famous by Polish anthropologist Bronislaw Malinowski, who studied them during World War I and wrote a classic monograph about the precocious sexuality of the islanders, *The Sexual Life of Savages*. He found it to be one of the few truly matrilineal societies, tracing ancestors only through the mother's side of the family, meaning that a man's children do not belong to his lineage, but his wife's. It sounded like a Christian missionary's nightmare, and for me a dream come true.

Getting there was another thing altogether. First I flew QANTAS to Cairns on the Great Barrier Reef in northeastern Australia. But the flight was four hours late arriving, so I missed the booked flight on Air Niugini to Port Moresby. Worse, my baggage was lost. There was, however, a second flight that day to Port Moresby and I managed to make it, only to discover the connecting Air Niugini domestic flight to Madang had been cancelled. Madang, on the north coast of Papua New Guinea, was where the *Melanesian Discoverer,* one of the world's largest luxury catamarans, was due to set sail for the Trobriands early the next day. Unfortunately, there were no flights to Madang until long after the boat's scheduled departure. I called the owner of the boat to beg a delay, but she had twenty Germans on board and said she couldn't hold the boat for me. I was on my own in New Guinea.

Opposite. The Trobriand Islands, often called "The Islands of Love," are famous for their raw South Seas sensuality and the precocious sexuality of the islanders.

The next day my wayward baggage caught up with me—it had been sightseeing in Sydney—and I discovered that a small New Guinea airline, Talair, actually had a once-a-week flight to a coral strip on Kiriwina, the main island in the Trobriands. By luck it was scheduled to depart the next day, so I signed on.

It was a mildly bumpy flight over the 3,000-meter peaks of the Owen Stanley range, then smooth sailing over the reef-studded waters of Milne Bay. There is little air traffic in this corridor, and nothing but ocean in all directions, so I wasn't too alarmed when I glanced up from the window to see our pilot, the sole cockpit occupant of our DeHavilland Twin Otter, slouched back in his chair reading the Papua New Guinea *Post-Courier.* The big black headlines screamed of impending war in the Gulf, and I knew at that moment there was no better place to be headed than the Islands of Love. Somehow the pilot managed to turn the last page just as we flew over Kaileuna, the first of the Trobriand atolls, and he turned to his instruments to begin our descent. There was a brief glide over blue water; then, as we floated above the edge of the main island, Kiriwina, I pressed my nose against the window to see a flock of sulphur-crested cockatoos burst from the mangroves.

The landing strip was half a century old, constructed by MacArthur's Seabees, and the terminal was nothing but a prefab tin hut. There are no phones in the Trobriands, but I had attempted a message by radio announcing my arrival. I had no idea whether the communication had been received or not. But sure enough I was met by Rebecca Young, the daughter of Speaker of the National Parliament Dennis Young, who owned the only hotel in the islands, the Kiriwina Lodge. Rebecca and her assistant, John Vaia, loaded my gear onto their Toyota, one of twelve motor vehicles on the islands, and we trundled eight

Below. The "Marys" dance naked from the waist up.
Opposite, top. A room with a view, from a Trobriand village hut.
Bottom left. The yam house in the village of Omarakana.
Bottom right. Tolabu takes a swim in a sinkhole.

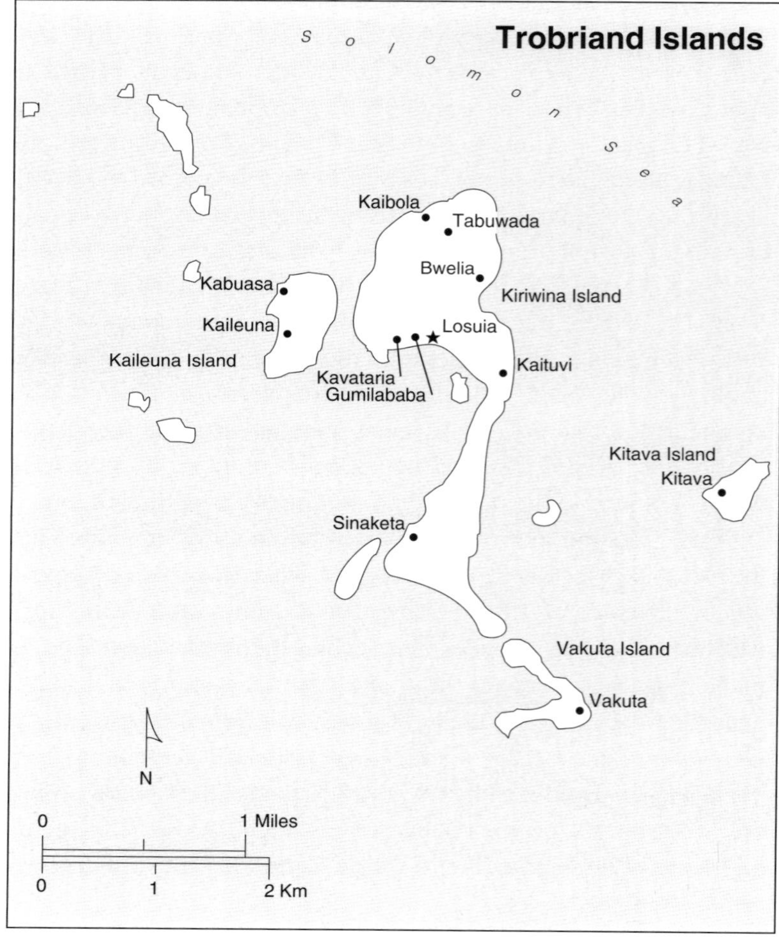

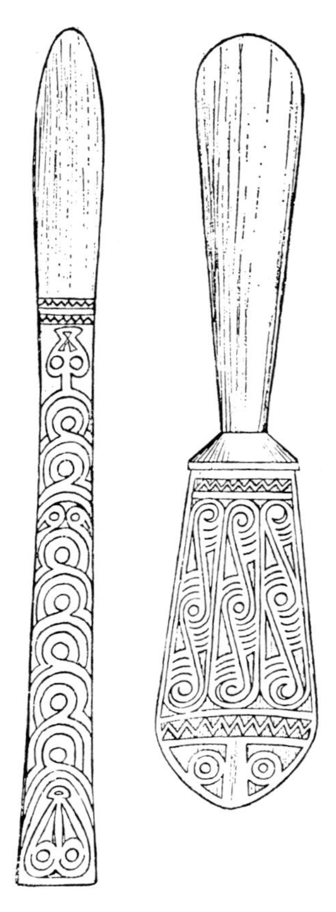

Wooden lime spatulas or *kena*. These spatulas are used for dipping lime from gourd storage containers—part of the betel-chewing process.

kilometers down a crushed coral road to the island's administrative and commercial center, the town of Losuia.

Jasmine, frangipani, and hibiscus trees marked the entrance to the lodge, but walking inside was like passing through a portal leaving the Garden of Eden. John led me through the dim lobby and past a warped pool table to my room. It was a disappointment, to say the least: dark, dank, with no air conditioning, no fan, no electricity, no towels, mouse droppings in the sink, and a small hard bed. A clacking gecko sat where the mint belonged. After the lizard scuttled out of sight, I asked John for a beer. He smiled apologetically. "Sorry . . . the bar doesn't open until 5:30, when we turn on the generator." I sighed. There seemed little paradisiacal or sensuous about this place . . . not even the price, $90 a night. This was where I had escaped, halfway across the world, to spend my fortieth birthday.

I wandered around the place until 5:30. This place was the pits, and in the back was something to prove it: a crocodile pit with three small crocodiles, one with fresh blood smeared across his back. Beyond the pit, still in sight of my room, was the hotel trash pile, at least a metric ton of crushed pop cans, beer bottles, food containers, and other hotel debris. Beyond that was a shallow and swampy bay.

Finally the hour was upon us. The diesel generator kicked into gear, the lights in the lobby came on, the bar cage rolled up, and it was time to celebrate. Three Australian guests joined me for an SP (South Pacific lager) and they wowed me with stories of the idyllic Islands of Love, since they had been here a full twenty-four hours longer than I. But their stories weren't of savage sensuality. In fact, they told me I'd missed the bacchanal Yam Festival by several weeks. The most impressive thing they'd discovered on the island was the cost. The hotel charged $1 per kilometer to drive one down the lone vehicular road that traverses Kiriwina; the food and beer were twice the price they were in Australia; and the carvings hawked by the locals started at $100. It was all a big ripoff, they complained.

Then a man came into the room, settled on the stool next to me, and ordered a gin and tonic. He knocked it back in a gulp, ordered another from the waitress, Jenny, and introduced himself to me as Bill Rudd, the manager of the Kiriwina Lodge.

Bill said he was the only permanent "tropo"—a white person gone tropical—in the Trobriands. The five other *dim-dims* (foreigners)—two missionaries, two anthropologists, and a school teacher—were on contract, and would return to their homelands at assignment's end. Bill looked as though he had been handsome once, but now, at fifty-two, with stooped shoulders, a mop of white hair, and the hooded, deep-blue eyes of a reptile, he looked like a character out of Conrad or Maugham, a refugee from the civilized world. And he was. Originally from Brisbane, he had come to New Guinea as a construction worker in 1962 when the country was still under Australian rule. Five years later he came to the Trobriands for holiday, and was so enamored of the lifestyle he vowed to return. He did, in 1970, and bought the lodge. But with so few tourists, less than two hundred a year on the average, he couldn't keep the business in the black. In 1985 he sold to Dennis Young, who in turn hired Bill as manager, as Dennis lived in Port Moresby. Bill was now married to Luisa, thirty-seven, his third wife and second New Guinean mate, and he had eight children, including two adopted, and Cocky, his pet cockatoo. He ordered another G&T and a couple of beers from Jenny, who was revealed to be his unwed seventeen-year-old daughter. She was pregnant, he joked, as a result of swimming in the lagoon. He ordered another round and told me more about himself and his adopted island. I drank with him, trying to keep up with his Falstaffian intake, and by dinnertime he offered to take me on a private tour of the island the next day—for only half price, 50 cents a kilometer. It was an offer I couldn't refuse.

Opposite, top. A Melanesian girl on a Trobriand beach with the *Melanesian Discoverer*, one of the world's largest luxury catamarans, moored offshore.
Bottom. The Kiriwina Lodge's crocodile pit.

Dinner was in an open-air room decorated with a single piece of local art: a large carving of a pair of pigs copulating. The fare was barely thawed, imported mystery meat ("John . . . is there any pepper to help spice up this meat?" "Sorry. No pepper."), boiled taro, yams, and pumpkin. The only drink available was instant coffee, and the sugar bowl was filled with red ants. When I'd had my fill I wandered back towards the bar. Bill was still knocking back G&Ts, so I veered to the right, found my clammy room, plopped on my bed, and began to read Malinowski's descriptions of the bizarre birth beliefs of the Trobrianders.

They believe that the spirits of the dead lead a happy life of perpetual youth on a small island called Tuma. However, sometimes they become bored and return, floating in on flotsam with the tide. They choose a woman of their own matrilineage and enter her body through the vagina as she bathes. Thus, girls who do not want to conceive avoid the ocean. Although today most Trobrianders know the biological facts of birth, they use these traditional myths to explain pregnancy out of wedlock or during a husband's absence. Since adultery is common but discovery dangerous, the concept of virgin-birth disguises inappropriate behavior, hence Bill's jest about Jenny's ill-fated swim in the lagoon.

Three different patterns of canoe prows.

The wind was swirling the next morning when Bill and his assistant, Tolabu, picked me up in front of the lodge. We were in the season of the southeasterlies, which didn't bode well. Bill's radio had reported that the *Melanesian Discoverer* had never left port as scheduled because the winds were so bad (so I could have made it on board if I had waited and taken the flight to Madang), and now it was several days behind schedule. It wouldn't reach the Trobriands until Friday, August 24, my birthday, if then.

Bill drove the Toyota south along the roadway built by the Allied forces in WW II. Along the way we passed dozens of strolling Trobrianders, their heads festooned with bright flowers. And we passed as many Ulysses butterflies, wings as blue and iridescent as a lava lamp.

Designs from lime spatulas and one small club (largest item).

The first stop was the village of Okaiboma, where Bill's *Kula* partner lived. *Kula* is a men's affair, a ritualized exchange of sea-shell ornaments during semiannual, interisland canoe expeditions. For a thousand generations necklaces and armbands have constantly circulated from one island within the Solomon Sea to another, never kept as personal possessions. The white armlets, called *mwali,* are made from cone shells and circulated counter-clockwise among the fixed ring of islands; the necklaces, called *bagi,* are made of hundreds of ground and polished discs from red Chama shells strung together by pandanus-tree fiber, and they circulate clockwise around the ceremonial trading circle, a cycle that often takes a decade. This elaborate trading ritual confers on the recipient an obligation to repay the gift at some future time; it also creates alliances among families and islands, and it is a vital social event. *Kula* partners are called *karayta'u,* and they are lifelong friends and allies, committed to offering hospitality, protection, and assistance to one another. Bill is the first white man to become a full-fledged member in good standing of the *Kula* Ring, and he is rightly proud of the honor. Here in Okaiboma Bill showed off a rack of his partner's *Kula* shells, including one very large piece named *Dogola,* dripping with cowrie pendants, string, and beads, and which Bill said was perhaps one hundred years old. Each piece is given a name upon completion, and the name is written upon it.

Here we ran into Peter Essex, a young photographer on assignment for *National Geographic* for a piece to be written by Paul Theroux. Peter said he'd made arrangements to live in one of the family huts because it was too expensive to stay at the Lodge, and I nodded with understanding. I wandered about the village for a few minutes, stepping over pigs, peering into the stilted homes with walls of plaited coconut leaves and sago palm roofs, and watching one woman stir a clay cooking pot filled with a yam-and-taro pudding. Another wove a sleeping mat with a fish-bone needle, and not far away a cluster of

Opposite, top. The traditional outrigger canoe.
Bottom. Each evening the canoes are parked in the elevated open individual stilt garages, just above the high tide mark.

Left. The Trobriands remain essentially unchanged since Denis de Trobriand, first lieutenant in the 1793 D'Entrecasteaux expedition, dropped anchor and left his name.
Above. A young girl anointed in coconut oil and flecked with yellow pollen prepares to dance.

Below. Enterprising Trobrianders paddle bananas and papayas out to the *Melanesian Discoverer.*

women stacked their money. (Here it literally grows on trees—their wealth includes the fiber of dried banana leaves). All this to the gay chatter of their soft-sounding language, an Austronesian dialect free of the Anglo missionary influences which had infected most other island language.

As we headed on down the areca palm-lined road Bill launched into one of his favorite topics, and one of the most discussed whenever the Trobriands are considered in academic circles. "Any primitive community that indulges in or is said to indulge in unrestricted sex behavior is considered an interesting community to hear from," the American anthropologist Professor Edward Sapir had written. Bill cheerfully agreed. "Sex here is like drinking water," he volunteered. "They're all sex addicts." He went on to describe how young Trobriander women are aggressive in seeking liaisons with lovers, how they are expected to have as many partners as possible from puberty until marriage, and how during the Yam Festival even the happily married go wild. He claimed that once during a Yam Festival he was walking down the road and was jumped by a group of young women who tore off his clothes and tried (unsuccessfully, he insisted), to gang rape him. He also claimed that, despite the rabid promiscuity, AIDS had yet to reach the Islands of Love.

Then Bill told me about his hero, King Cam. He was the uncrowned white monarch who ruled with absolute authority for fifty years on the easternmost Trobriand island, Kitava. He was the lord of a jealously guarded paradise where no other white man was allowed to set foot, and he was the master of his own harem. His real name was Cyril Bernevelot Cameron. The black sheep of a wealthy Scots family, he was born in Australia

in 1887. Around 1909 he turned up in New Guinea and went fossicking for gold in the wild interior. When he fell in love with the beautiful wife of a white planter and was rejected, he bought a schooner and set out to establish a copra plantation on the remotest island he could find. Chance took him to the pear-shaped island of Kitava, where the big, rawboned Scot with piercing green-blue eyes pitched a tent on the beach. It wasn't long before the natives built him a house and helped him establish his coconut plantation—and his reputation. His word soon became their law. Trading schooners called occasionally to pick up Cameron's copra, but no white man ever ventured beyond the coral foreshore. If sailors lingered King Cam reminded them of his loaded Colt revolver on the table beside the rum glasses in his palace. Presaging future military euphemisms, he called the Colt his Peacemaker.

Until the day King Cam died in 1966 at the age of seventy-eight, no male native was ever permitted to set foot inside his palm-thatched palace; only the women, many of whom became his wives, entered Cam's castle. Legend had it he kept one hundred girls in his harem, that he would begin to train them at age twelve, then dismiss them at fourteen. "Bloody wooden-headed niggers" was how King Cam described his loyal subjects, but he apparently cared for them nonetheless: he left the Kitavans every penny of his $200,000 fortune. He died in a hospital in Port Moresby, the last man to rule a private island kingdom in the Pacific, and the living realization of many a modern male's frustrated ambitions, including Bill Rudd's.

Dance shield—a rubbing of the pattern.

After we'd bounced along and listened to Bill's stories for a couple of hours the neighborhood changed. With little transition we were in the midst of a tropical rain forest. Creepers hung in curtains from the tall strangler figs and tropical hardwoods crowded either side of the road. But the ride didn't go as planned. Some five kilometers from the southern tip of Kiriwina, we were stopped by a large, fallen banyan tree. Someone had tried to burn through its base, and thin curls of smoke were still rising, but the fire had extinguished itself halfway through the stump. There was no way to move this mammoth tree, and it would take hours to cut through, so we decided to get out and walk a spell.

The first dividend of this decision was the birdlife. The sky was cleaved with screeching eclectus parrots, Eastern black-capped lorikeets, and sulphur-crested cockatoos. This was a rain forest as untouched as the middle Amazon, a garden untilled by thinking creatures, and it beckoned like an exquisite symphony.

We walked just a short distance before reaching the next reward: a small path led from the road to Lukwevesa, an inviting sinkhole the size of a suburban basement. The island is honeycombed with natural cisterns, and from these much of the fresh water is gathered for the sixty villages and 22,000 inhabitants of Kiriwina, as there are no rivers or streams. But this sinkhole, far from any village, was born to be swum. Within minutes of arriving, Bill, Tolabu, and I were stripped and diving into the clear, cool waters of this subterranean natatorium. It was the most refreshing moment of my journey so far, and as we dogpaddled and explored its nooks and niches a sea eagle swam in the sky above.

On the way back to the lodge we stopped at the island's primary school. Young girls, crowns of red hibiscus blossoming in their shiny black hair, were out in front learning a dance in which they swung their traditional grass skirts with lissome grace while clapping coconut halves. The boys were in the back practicing something called the Snake Dance. Even though prepubescent, both camps were a mass of writhing and swirling hips, much like an MTV dance party.

The segregated nature of the dances said as much about the peculiar mores and ambiguity of these islanders as anything. By the time children are seven or eight years old they have begun playing erotic games with each other and imitating adult seductive attitudes. Then, although they begin to *kayta* (have intercourse) very soon after puberty (and

sometimes before), and thereafter pursue sex avidly, changing partners often, the rules of modesty prevail. Four or five couples may be using the *bukumatula* (bachelor house) on a given night, but one should not look at the other couples, and during the daylight there are no public displays of affection. Perhaps most important, lovers never eat together. "Europeans object to an unmarried woman sharing a man's bed," wrote Malinowski, "while the Trobriander would object just as strongly to her sharing his meal." Only as husband and wife can they eat together, and, in fact, there is no marriage ceremony other than the girl and boy openly sharing a meal in the home of his parents. Once married, monogamy is the rule, except during the Yam Festival, when all rules are abandoned, and women of all ages and statuses seek intimacies with the men. It's sort of Sadie Hawkins Day in the extreme.

The longtailed bird of paradise.

The next morning, August 24, we decided to explore the north, with hopes the *Melanesian Discoverer* would arrive at Kaibola on the north coast. First we stopped at the village of Omarakana, the home of the island's Paramount Chief and his wives, and thus the largest yam house on the island, one that dwarfed the surrounding dwellings. It was a magnificent piece of architecture, painted in the favorite regional colors of red, black, and white, and decorated with an elaborate painting within its gables. And it was filled with yams, collected and stored during the recent harvest. We missed Pulayasi, the sixty-year-old chief, who was off on some island chore, but I was shown his latest treasure, Toimatananoli, an enormous *Kula* shell he found the week before on Kitava. I was also shown the site where Bronislaw Malinowski had pitched his tent during one of his ethnographic visits.

By late afternoon we arrived at Kaibola, where I witnessed the most gorgeous sight during my stay on Kiriwina: the white-hulled *Melanesian Discoverer* anchored off shore, and two yellow Lancer inflatables rocking gently against the crescent beach, suggesting wonderful things ahead. A score of white-skinned tourists were bargaining with the scores of brown-skinned entrepreneurs hawking their wares: erotic ebony carvings, stylized walking sticks, *kwila* bowls polished to a high gloss with boar's tusk, crescents of pearl shell, grass skirts, tortoise-shell earrings, and cowrie necklaces. Here I also met Trevor, an Australian from Adelaide who has been coming to Kaibola beach for holiday for seven years. He told me he had a hut built on the water's edge, and for six weeks each year he would come to Kiriwina and fulfill his fantasies. Then he gave me a sly, smug smile, as though he alone knew the passage to Shangri-la, and he wasn't going to share his secret with a soul.

Papuan pipe.

Trevor disappeared into the trees just as the boat's German tour leader motioned to come watch the famous dances. As we followed a procession of Trobriand women towards the village soccer field I couldn't help but notice their warm-up apparel. One was sporting nothing up top but a Maidenform bra; another wore a T-shirt showing a small, smirking cat with the legend MEET A HAPPY PUSSY; another shirt proclaimed, BEAUTY IS OUR LAND; PARADISE IS THE EXPERIENCE, and yet another woman was carrying a Gucci purse. The first beguine, performed in front of the goalposts, featured the lissome "Marys" (the pidgin term for native women), naked from the waist up, wearing gaudy, three-colored, banana-fiber miniskirts, undulating and swaying in a vaguely snakelike fashion. Their nut-brown naked skin was anointed in coconut oil and flecked with yellow pollen, and in their thick hair they wore black cassowary and white cockatoo feathers. When the *kundu* drums stopped they all got together for a lengthy, laugh-filled tug-o-war. As the show was going on I was struck by the contrast in cultures—the color and open glee of the Trobrianders, the drab khaki dress and serious faces of the Westerners. There was one exception. A woman on the sidelines, a Caucasian with a wide smile, the eyes of Salome, and a silver streak fanning like a shooting star through auburn hair, was photographing the events. She was intent on the goings-on, but occasionally would look up and send a quick

Opposite, top. The men perform the Tapioca dance, a Melanesian lambada. *Bottom.* By the time Trobriand children are eight they have begun playing erotic games and imitating adult seductive attitudes, and they begin having sex soon after puberty.

glance over at me. The Trobrianders yelped and giggled, their breasts bouncing, and I kept noticing the woman photographer, who in turn kept noticing me.

Then the men marched out of the forest, wearing tight, orchid-fiber arm bands and white codpieces fashioned from strips of bark. They performed something called the Tapioca dance, in which they thrusted pelvises and slapped their own fannies, as though simulating sex standing up. It was like a Melanesian lambada, and as they circled the arena in a sort of conga line they chanted between the intermittent shrieks of a whistle blown by the leader. This was a dance that, when performed during the Yam Festival, is a preamble to a night of indiscriminate sexuality. This was the off-season, but it still seemed daring and erotic, and at one point the line of men and boys started stomping towards the pale-skinned photographer with the ray of silver in her hair. She pulled the Canon from her face and arched her eyebrows as they made overtures. Then she looked over and hurled a mischievous grin in my direction. The air was thick with first-date humidity. It was near the end of a hot day, but the temperature seemed to rise.

A nineteenth-century Trobriander.

When at last it was time to go I bade farewell to King Bill and boarded the yellow life raft for the trip across a stretch of translucent sea to the sleek, ultramodern, thirty-six-meter-long *Melanesian Discoverer.* It was sunset as we motored back, and in the distance I could see the silhouettes of the spreading matting of pandanus boats, the woven sails billowing like butterfly wings. After I had a much-needed shower, a four-course dinner of fresh mackerel and crisp vegetables was served on bone china in the air-conditioned lounge. It was heaven on water. Afterward Jan Barter, the owner of the boat, announced that we should all head outside to the aft deck. There we watched as a dozen splinter-thin outrigger canoes paddled up and moored to the back of the *Discoverer,* and young women and men dressed in ceremonial finery climbed on deck. The boys began strumming guitars as a bevy of small yet perfectly formed Trobriander girls began a fertility dance in their beautifully decorated and very short fibrous skirts. There was something exotically reptilian and not quite human in their movements. It was the most concupiscent dance yet. I clapped and admired, and looked around at the other guests, stopping at the attractive silver-streaked woman. She was applauding as well, and routing a furtive, provocative smile towards me. Then the Trobrianders asked the *dim-dims* to join in the dances. I was pulled out to the middle of the deck by a glistening, perfumed, sixteen-year-old beauty with bare breasts and a ring of powdered coral lime decorating an almond eye. I tried to duplicate her hip movements with little success. Then another bare-breasted maiden enticed the photographer up to the dancing circle, and suddenly she was shimmying alongside me. Every now and then she would brush her hip against mine, and look over, her face creased in a beatific smile as though suggesting this was just the prelude to a better dance. I grew steadily more dreamy with desire as I did a hula to the throbbing beat of a lizard-skin drum. This Melanesian mambo went on for hours, and was still going on when a wave of exhaustion washed through me. I bid Jan Barter goodnight, and turned to say the same to the silver-streaked woman, but she was gone.

As I walked down the deck to my stateroom, I could hear a gentle lapping against the boat, and the soft murmur of the sea. Or was it the sea? I looked up and saw a single star shining limpid and tremulous, like a dewdrop about to fall. Then I remembered that today had been my birthday. It had been a pretty good birthday, I acknowledged to myself, though perhaps not what I had fantasied.

Then I opened the stateroom door and looked across a dusky room. On my bed was the woman with the shooting star in her hair, flashing that familiar smile. On the table was a bottle of Dom Perignon, the liquid franca of choice, and a couple of colored balloons. As I leaned across the bed to give her a kiss, she stretched her arms around me and whispered, "Happy Birthday, honey. It's great being married to you."

Easter Island

Exploring An Elusive Yesterday

For the whole air vibrates with a vast purpose and energy which has been and is no more.
—KATHERINE SCORESBY ROUTLEDGE,
The Mystery of Easter Island: The Story of an Expedition, 1919

The women (of Easter Island) are not quite disagreeable.
—JOHN REINHOLD FORSTER, *Observations Made During a Voyage Round the World,* 1778

She rode bareback along the wind-whipped coast. Mind you, the horse wasn't bareback, she was. She'd gone native, a Polynesian-style Lady Godiva, here on the most solitary inhabited spit of land in the world . . . Easter Island.

A flyspeck on the map, only eight miles wide and fourteen miles long, its rounded deltoid-shaped surface is about the size of San Francisco. It sits at 27 degrees south of the equator, fretted by incessant winds, ringed by a million square miles of open sea. Motherland Chile lies over 2,300 miles to the east. The closest inhabited place, almost 1,200 miles to the west, is Pitcairn Island, settled in 1790 by mutineers from the H.M.S. *Bounty.* To the north, a vast desert of Pacific Ocean; to the south, the Austral Seas as far as the glacial coasts of Antarctica. The nearest solid land the islanders can see is the moon.

Its landscape is as bleak as the island is remote—buffered by distance, as denuded as a lunar landing site, as naked as Deborah's back as she trotted toward a towering stone figure with no eyes.

Its prehistoric residents called it Te Pito-o-Te Henua, meaning "The Navel of the Earth." The Tahitians, who arrived during the nineteenth century, called the island Rapa Nui, a Polynesian word for "Big Paddle," a needed tool to get here. But when Dutch explorer Admiral Jakob Roggeveen dropped anchor on the eve of Easter Sunday in 1722, its current cartographic celebrity was born. Half a century after the Dutch, the Spaniards arrived for a brief stay in 1770 under orders from King Carlos III. They found little of interest on the lonesome, raw-boned isle, but they renamed it San Carlos, and half-heartedly annexed it for the Spanish Crown.

Four years later Captain James Cook sailed in to replenish his stock of food and water. There was little of either. The English were amazed at the giant, flat-headed statues with their inscrutable sphinx-like faces and cylindrical ten-ton topknots of red stone balancing on their brows. "It was incomprehensible to me," Cook wrote, "how such great masses could be formed by a set of people among whom we saw no tools."

In 1888 the Chilean government finally annexed the island, without protest from Spain or any other country. As Captain Cook had said, "No nation need contend for the honor of the discovery of this island."

Opposite. With over 600 lava-hewn statues and untold thousands of ruins and artifacts, Easter Island is the richest, most bewitching open-air museum in existence.

Below. Moai are recognized throughout the world as the dour, unblinking symbols of Easter Island.
Opposite, top. In no other place on the globe can such large statues be found in such a small place in such great numbers.
Bottom. Applying tattoo paint for the annual Tapati Rapa Nui festival.

The archeological history of the island is as incongruous as its name. Shrouded in tantalizing mysteries and etiological puzzles, the place continues to pique the minds of scholars. With over six hundred lava-hewn statues and untold thousands of ruins and artifacts, most dating to pre-thirteenth century, the island is the richest, most bewitching open-air museum in existence. The colossi of rock, *Moai,* as the islanders call them, are recognized throughout the world as the dour, unblinking symbols of Easter Island, though relatively few outsiders have seen them in the stone. In no other place on the globe can such large statues be found in such a small place in such great numbers and created by so few people. They stand scattered like pawns in a chess game of the gods, played against the background of the ocean and the ages, waiting for the next move of a long-hesitant hand.

I arrived from Tahiti with Deborah Dunston, an eco-anthropologist from the University of Guam. She was here to see if she could debunk some of the common theories about why the ancient cultures who carved these elongated stone heads disappeared. I was here to watch.

After a breakfast of hard biscuits and two cups of Nescafe Instant (the well-appointed hotel insisted no real coffee was available on the island—too expensive to import), we shouldered rucksacks with lunch and camera gear and wandered outside to our two—ahem—horses. Not quite black stallions. Some 4,000 wild horses roam the island, outnumbering the human population by a thousand, and the horse is the most common mode of transportation. We were blessed with two relatively tame specimens, and they looked it: haggard, dull coats, sagging middles, and tired eyes. But the price was right ($10 a day), so we mounted the gunny-sack and sheep-fleece saddles and with a crude map in hand started loping northwards up the jagged coast.

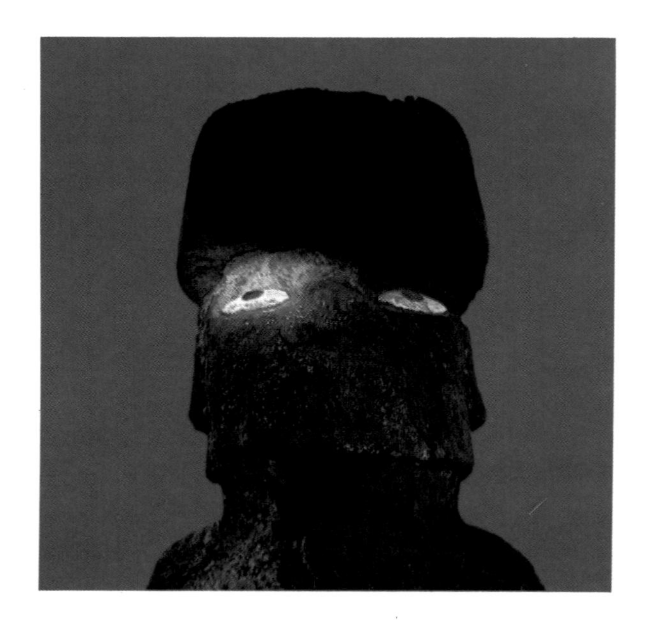

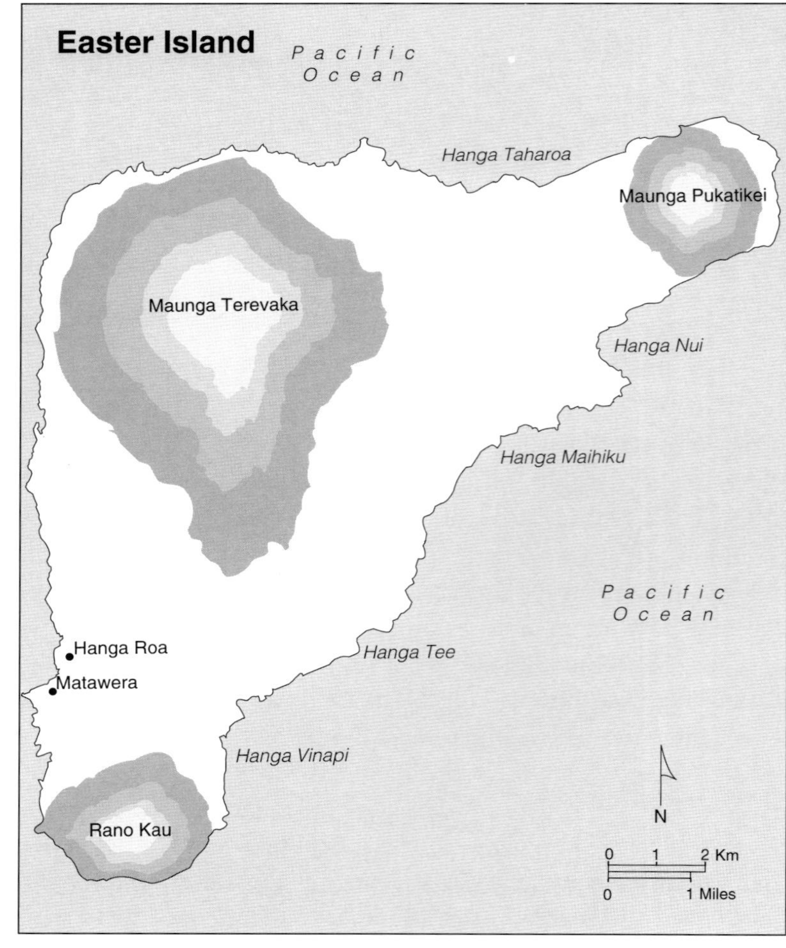

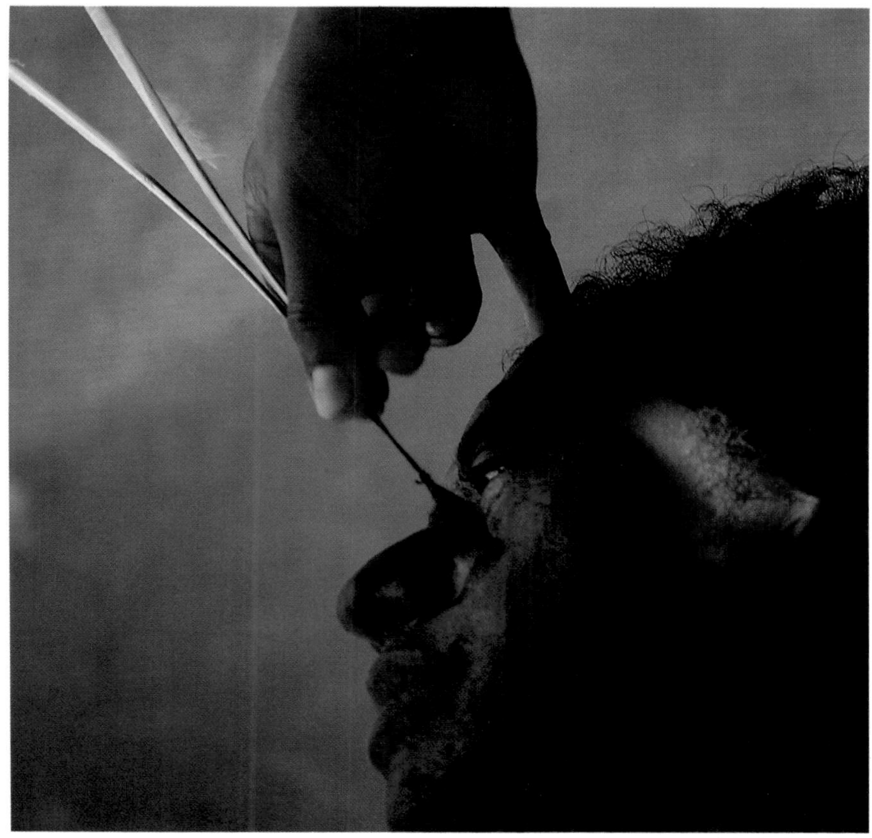

Opposite, top. Most of the Easter Island statues were made from volcanic rock found in the tuff quarries of Rana Raraku.
Bottom. Some 4,000 wild horses roam the island, outnumbering the human population by at least 1,000.

We first passed through the island's single settlement of note, Hanga Roa, with its tidy white houses of cinder block or painted wood and tin roofs that served to collect water. Stopping at the bleached stone library we asked if we could borrow the island's only copy of *AKU-AKU*, Thor Heyerdahl's famous account of his 1955 archeological expedition. "Sorry, no lending," the librarian said in Spanish. Just as well, perhaps, as an archeological student volunteered that in his opinion the book was "more fiction than fact." And, in fact, the controversies still rage as hard as the sea into the cliffs on this three-million-year-old volcanic mote. After six expeditions to Easter Island, including one in 1986 during which he uncovered ancient ceremonial constructions, human teeth and bones, soil containing charcoal, and stonework similar to that found in pre-Inca sites in South America, Thor Heyerdahl stoutly believes the island's first inhabitants arrived from Peru about 800 B.C., travelling in balsa-log rafts such as the one he sailed from Peru to Polynesia in 1947, the raft he called Kon Tiki. But most of the Pascuenses, as the islanders call themselves, cling to the belief that their ancestors came from Polynesia. Many reputable archaeologists and other scientists agree, refuting the "daring Viking's" notions as self-serving and romantic. Those who take issue with Heyerdahl believe the evidence is strong that Easter Islanders settled around A.D. 400, arriving from islands within Oceania. Besides similar racial and language characteristics, they cite ancient carved symbols on the island also found on art objects in other parts of Polynesia.

Less than three miles up the coast we cantered across a fawn-colored meadow, flushing hawks to flight. Then we crested a ridge and stopped in our tracks. There, out of the dusk of centuries, silhouetted against a shimmering expanse of sea, looming above us, were the five statues called Ahu Tahai. Deborah rode to the base of these immense, mute monuments. Arching her neck, she locked eyes with the hollow sockets of one stone sentry, and the brooding conundrums of Easter Island stormed with the whirling wind.

More impressive than Stonehenge or Carnac, the Easter Island statues have spawned theories of origins from the plausible to the preposterous: architects from the lost city of Atlantis, ancient astronauts from deep space, South Sea spirits that walked onto the beach and froze, the art of Incan sailors, and more. Not one of the Easter Island faces is smiling, so Deborah decided this was evidence the chiselers were originally from Oakland.

An Easter Island woman, sketched during a voyage of Captain Cook, 1774. Islanders divide themselves into descendants of two distinct ethnic groups, "long ears" and "short ears." This woman was one of the "long ears."

The statues range from nine to sixty feet in height and from three to one hundred tons in weight, and are scattered around the island in a pell-mell fashion. There are virtually no trees on Easter Island, so there's no wood to fashion into wheels or rollers or levers. It seems incomprehensible that a people who had no natural means of leverage or traction could make, lift, and move these creations without supernatural or extraterrestrial assistance.

Where did these people come from, where did they go, and how did they move these statues?

Thor Heyerdahl believes the carvers came from South America. The resemblance to ancient structures in Cuzco, Peru is uncanny, and *Cuzco* in the Incan language means "navel," as in the islanders' original name for their home.

But the more accepted ethnographic explanation depicts a migration from the Marquesas Archipelago bringing, in great outrigger canoes, a population of gifted sculptors to the island, perhaps as early as the seventh century. Oral tradition tells us that the statues were intended as tributes to ancestors, and that ingenious methods were devised to transport and raise them. Most were made from volcanic rock found in the tuff quarries of Rano Raraku, on the southeastern coast, and were fashioned by stone chisels made of obsidian.

A second migration from Mangareva in the Gambier Islands, 1,700 miles to the north, brought a less creative, more agriculturally inclined group, who, it has been postulated, at some point clashed with the Marquesas emigrants and wiped them out.

There is another, more recent, theory as to why people stopped carving these incredible statues, one that Deborah Dunston endorsed. It concerns damage to the environment, and it is a theory that upsets some anthropological applecarts. After all, it is an accepted cliché that primitive peoples lived in perfect harmony with the land, taking only what they needed and causing no harm. But anthropologists are discovering that in fact a number of ancient cultures—such as the Anasazi of the American Southwest—were nearly as adept as modern humanity at ecological destruction. Although it is impossible to prove direct cause and effect, scientists are learning that the collapse of a number of cultures took place just after major human-caused ecological changes.

Unless some mysterious spiritual revelation abruptly stopped the sculpturing, it seems the culture that produced the statues of Easter Island came to an untimely end. Perhaps it was a civil war. Maybe a catastrophic eruption from one of the island's volcanoes. But John Flenley and Sarah King of the University of Hull in England have another hypothesis. "What we have done is investigate samples of fossilized pollen from crater swamps on the island," Flenley wrote in the British journal *Nature*. The pollen record is remarkably complete, documenting the existence of over forty plant species, including giant palms, and offering a detailed look at how the island's vegetation changed.

Stone figure.

Mural painting.

The oldest pollen dates back some 30,000 years, long before the first people arrived, indicating the island was then heavily forested. Whenever the island was finally settled, its inhabitants began to clear the land for agriculture and livestock. They also cut trees to build canoes, and eventually logs for transporting and erecting the statues. The land was relatively fertile, the sea teemed with fish, and the people flourished. The population rose to about 15,000, and the culture grew sophisticated enough to carve the giant statues.

Unfortunately, the trees didn't grow back. "The pollen record shows the deforestation beautifully," says Flenley. "It began about 1,200 years ago and was almost complete by 800 years ago." The islanders also exploited many of the island's other resources, such as its abundant bird eggs.

The result was an ecological disaster. The Easter Islanders had cleared so much of the forest that they were without trees to build canoes for fishing. They also had probably taken so many eggs of the sooty tern that the bird no longer nested on the island. To make matters worse, deforestation caused irreversible soil erosion, destroyed the topsoil, and severely reduced crop yields. Fewer fish, eggs, and crops led to a shortage of food. Hunger, in turn, brought warfare, even cannibalism, and pushed the whole civilization over the brink of collapse. By the time European explorers arrived the population was down to about 4,000, and the culture that had produced the statues had disappeared entirely.

For Professor Flenley the moral could not be clearer: "Easter Island is the world in miniature. What happened there is what we are doing now to our rain forests and our other renewable resources."

After a volley of photographs, Deborah and I continued up the barren coast a short ways, and stopped at a bluff to watch the sea fling veils of water a hundred feet into the air. At this meeting of rock, sea, and sky—mass, energy, and light—Deborah felt inspired to pull off her shirt, as though there might be merit to meeting the elements, and the spirits of the island, in her most candid state. Besides, she said, it was quite hot. Turning inland, we switched our steeds to move them out of "poke." Mine, inappropriately enough, was named Pegasus. In short order we were at the base of Ahu Tepeu, a magnificent beetle-browed statue crowned with a red stone headdress weighing eleven tons. The achievement of donning this fellow's hat must be compared with putting a man on Mars today. The best of origin theories notwithstanding, the erectors likely had little wood at

Opposite, top. Easter Island sits at 27 degrees south of the equator, fretted by incessant winds and ringed by a million square miles of open sea.
Bottom. It is the most solitary inhabited spit of land in the world. The nearest solid land the islanders can see is the moon.

Left. When Dutch explorer Admiral Jakob Roggeveen dropped anchor on the eve of Easter Sunday in 1722, the island's current cartological celebrity was born.
Below. When Captain Cook landed in 1744 he was amazed at the giant, flat-headed statues that towered over him with ten-ton topknots of red stone balancing on their brows.

Opposite, top. The statues are scattered like pawns in a chess game for the gods. *Bottom left.* How did the carvers move these giant statues? There are virtually no trees on Easter Island, so no wood to fashion into wheels or rollers or levers. *Bottom right.* The tattoos of today are actually dyes made from colored clay.

their disposal and limited manpower; but the statue stands, proud in his haberdashery, lips mysteriously pursed, eyes blind, mouth in solemn silence, yet somehow alive in the deadness of stone.

Ahu Tepeu faced inland, as do almost all the statues on Easter Island. A popular theory is that the statues were created to represent important people who had died. The power of the deceased was thought to be transmitted to descendants through the eyes of *moai*. Thus, all the statues originally faced the center of the island, toward the villages. Deborah guided her horse behind the statue while gaping at the huge hat, and suddenly her indolent animal reared, almost knocking Deborah off. I dug my heels into Pegasus, and he reluctantly circled to the back side as well. When he looked up at the statue, he also reared and whinnied, almost tossing me to the dirt. Looking up, I saw the source of the animals' fright—from this vantage it appeared the statue was toppling over towards us, a strong illusion that matched the spooky nature of the place.

For the next two hours the ride yielded nothing, save stark vistas, a rough pitchstone terrain, and wild horses. The island is entirely volcanic, with three major cones forming the points of a triangle. As we zigzagged northwards we found ourselves ascending the talus slopes of the island's highest peak, the extinct Volcan Aroi, 1,400 feet above the sea. Halfway up, a curious grove of banana trees circumscribed a rock outcropping and Deborah dismounted to investigate.

"Richard, there's a cave in here." She spoke in the hushed tones expected in a haunted castle.

The island is honeycombed with lava caves, and Deborah had stumbled upon a large one. She poked her head inside, waited for eyes to adjust, and then squealed in a muffled voice, "There's a skull of a ram inside. Can you try to get to it?"

A boulder blocked the entrance, but with both backs into it we were able to roll it aside. A shaft of light struck the horned skull and sent a shiver through my body.

Drawing of an Easter Island canoe from a voyage of Captain Cook, 1774. "The boat which came off seemed to be a very wretched thing, patched together of several pieces, the head & stern high & the middle low; there was an outrigger fixed to it, & each of the men had a paddle made of more than one piece, which sufficiently proves the want of wood in this isle."
—Johann Reinhold Forster, 14 March 1774

I lowered myself into the grotto feet first, kicking aside a latticework of spiderwebs. Inside I squirmed to my knees and crawled through the damp, black velvet of darkness to the skull, which was lit by a pinpoint of sunlight. Next to it, in the half light, I could make out two more skulls. I reached to pull one closer, then coiled back like a snake-bitten dog.

"Deborah, there are two human skulls in here," I said shakily. I brought the skulls to the surface to photograph, and saw that each had a pen-sized hole in one side of the head, and a jagged, gaping grapefruit-sized hole on the other. Forensics is hardly my forte, but the marks looked like bullet holes to me. The eeriness of Easter Island seemed most acute at that moment. What chilling stories would the heads tell could they speak? Murder? Accident? Cannibalism? Double suicide? How old were they? One month, one year, one hundred? Did they know answers to the great riddles of the island?

Later, back in Hanga Roa, we spoke with Claudio Cristino, an archeologist from the University of Chile who had spent the last four years studying and mapping the island's thousands of archeological sites.

"Those caves are sepulchers, burials chambers for the victims of smallpox back in the mid-1800s," he told us.

Claudio agreed with Professors Flenley and King that Easter Island at its height supported 15,000 people, a bustling South Pacific station. When Captain Cook arrived he found only six hundred men and fewer than thirty women eking out existences on an island with only stunted mulberries and tiny mimosas for trees. "On the entire surface of the island, there is not a tree that merits being called that," wrote naturalist George Forster, who accompanied Captain Cook. If the ecological devastation theory holds, most of the population loss was a result of forest obliteration more than six hundred years before

Opposite. Some scholars believe Easter Island was once heavily forested, but when its ancient inhabitants cleared the land for agriculture and livestock, the trees didn't grow back.
Above. Climbing Mount Puna Pau, a dormant volcano behind the village of Hanga Roa.
Left. Some say surfing was invented on Easter Island. The thick but lightweight rush weeds called *totora* are woven into conically shaped floats.

Cook's landing. But things got worse. In the early nineteenth century Peruvian expeditioneers, looking for cheap labor, abducted Easter Islanders as slaves, and introduced smallpox, consumption, and venereal diseases to those remaining. By the mid-nineteenth century the island's population was decimated. At its ebb, in the 1870s, there were just 111 inhabitants. Today, almost a century and a quarter later, the population has grown to over 3,000, and the place still seems underpopulated.

After our skullduggery at the cave we spurred the horses onward and upwards. We came upon a simple farmhouse, an island of life on the desolate volcanic slope, where a dark, disheveled figure stepped out to meet us. As he approached, Deborah, in a moment of modesty, turned away and pulled her shirt back on. Then she turned back to face the farmer, and dropped her mouth in shock. The left side of the farmer's face was grotesquely contorted in bizarre lines, with lip and eye dropping like melted butter. He was a leper, one of about thirty on Easter Island, and his disease had paralyzed and disfigured his face. Now he lived in isolation on the world's most isolated isle.

When Chilean navigator Captain Policarpo Toro negotiated to transfer Easter Island to Chilean sovereignty in 1888, he brought with him several Easter Islanders who had been living in Tahiti. Missionary records indicate that one passenger was visibly ill with leprosy, already showing some limb paralysis. He was the first.

The disease spread quickly and a decade later a leper colony was built not far from this farmhouse to isolate the sufferers. By the 1940s, forty islanders had the disease. Then, with island-wide vaccinations in the '60s and '70s, the disease was at last officially eradicated. Now the last of the lepers have staked out homesteads in the far corners of the island, such as the one here on the side of the volcano.

Below left. The traditional form of transport across this San Francisco-sized island is by horse.
Right. Some line fishermen still use hooks carved from stone and bone, a tradition dating back centuries.

We nodded and tried to exchange salutations, but were hampered by the impenetrability of a native dialect we couldn't understand. He smiled, and waved us towards his home, so we slipped off the horses and followed him inside. There he pulled a black pot off the stove and served us cups of steaming, delicious, real bean coffee. It was an unexpected treat, and when I asked him in my best sign-language what we could give him in return, he shook his head. I insisted, and finally, after some thought, he tugged at Deborah's shirt with a wiry arm. We all smiled, and I pulled my Hanes T-shirt off instead and handed it to our host.

Easter Island idol.

After bidding goodbye to our new friend we continued the ride up the fallow grade, reaching the summit midafternoon. A shallow crater, lush with rain-nourished grass (the island is devoid of running water) formed an imperfect crown. Some of this grass was papyrus, known as *totora,* like that found along the shores of Lake Titicaca, and the stuff Heyerdahl believes made up ancient ocean crafts. Though the last eruption was over 2,000 years ago, we couldn't help but envision a Mount St. Helens scenario as we rimmed the crater on horseback, drinking in the glazed Pacific horizons at every point of the compass.

The horses picked up speed and fire as we descended the eastern scree slope for the return journey. After an hour's hard ride we crested an empty ridge and looked down upon Easter Island's most resplendent sight—Ahu Akivi or "The Seven Monkeys," as the islanders have irreverently nicknamed them. Since restoration by Chilean archeologist Dr. Gonzalo Figueroa and the late Professor William Mulloy, former head of the Department of Archeology at the University of Wyoming, the seven monkeys have become the most renowned and most photographed statues on the island. They stand like soldiers guarding a wasteland, fixed in scorn, forever watching a vacant landscape and the watery azimuth beyond. Their graven visages serve as tongue-tied testimony to a past about which we can only surmise and quarrel.

Minutes later our once-slow-as-glue horses were galloping back Preakness-style, Deborah and I clinging like cats to their backs. Minus my right stirrup we screeched into Hanga Roa, pulled into the first tavern, wrapped the reins around a hitching post, and moseyed inside for a brew. We ordered a Brazilian import called Xingu, and watched as several local women boogied to the strains of the Bee Gees' "Night Fever."

I walked outside and gave Pegasus a stroke, and pulled the fleece saddle off his sweaty back. A gust of wind spun down the lane and pitched dust into my eyes. A chill ran through me. I still had no shirt, having left mine with the leper on the hill. But the chill was from more than weather. Perhaps the ghosts of sculptors past were warning of the dangers of not being careful stewards of our land and resources. If primitive people were capable of creating great art, and then destroying their culture and civilization through destruction of the environment, where are we, their ignoble and industrialized progeny, headed? We have one great advantage. We can learn from the lessons left behind, the messages conveyed in the stony stares of the sentries of centuries past.

I took a long draw from my Xingu beer, and stared into the splendidly lonely landscape. Not far away the sea was murmuring something about hope. The sun was setting, but I imagined I could see a slight, sly smile on the lips of the statues on the ridge.

Other Exotic Islands of the World

Aleutians

A chain of rugged, volcanic islands curving 1,200 miles west from the tip of the Alaska Peninsula and approaching the Komandorski Islands of the Soviet Union. A partially submerged continuation of the Aleutian Range, they separate the Bering Sea from the Pacific Ocean. The Aleutians are composed of four main groups: Fox Islands, nearest to the mainland; Andreanof Islands; Rat Islands; and Near Islands, the smallest and westernmost group. The Aleutians have few good harbors, and the numerous reefs make navigation treacherous. Temperatures are relatively moderate, but heavy rains and constant fog make the climate dreary. Almost completely treeless, the islands have a lxuriant growth of grasses, bushes, and sedges. Sheep and reindeer are raised. Hunting and fishing are the main occupation of the islands' Eskimo population. The Aleutians were discovered in 1741 by Vitus Bering, a Danish explorer employed by Russia. The indigenous Aleuts were exploited by the Russian trappers and traders who, in search of sea otter, seal, and fox fur, established settlements on the islands in the late eighteenth and early nineteenth centuries. The Aleutians were included in the Alaska purchase in 1867, and at that time became part of the U.S. Dutch Harbor, one of the few good harbors, became a transhipping point for Nome in 1900, after the discovery of gold turned Nome into a boom town. During World War II a U.S. naval base was established at Dutch Harbor. In 1942 the Japanese bombed the base and later occupied Attu, Kiska and Agattu islands. The U.S. regained the islands in 1943. Now, the area is one of the last true wilderness island chains in the world.

Andaman and Nicobar Islands

This is home to India's finest undersea park, the Wandoor National Marine Park, comprising twelve islands lapped by crystal-clear waters, and famed for its tropical fish and extraordinary coral formations. The islands, known to Europeans since the seventh century A.D., sitting in the eastern Bay of Bengal, consist of more than 200 land specks off the coast of Burma. They were the site of a British penal colony from 1858 to 1945. In the '50s the Andamans were colonized with displaced persons from East Pakistan (now Bangladesh), evacuees from Burma, and Indian emigrants from British Guiana (now Guyana). Today the islands are a union territory of the Republic of India. Much of the surface is covered with rich, dense forests, including redwoods and gurjan. The islands are extremely remote, and portions are still inhabited by hunter-gatherer tribes that have had no exposure to modern civilization. The islands are restricted territories, and special permits are required for a visit. A number of the islands, including all of the Nicobar group, remain off-limits to visitors. The capital of both groups is Port Blair on South Andaman Island.

Opposite. The Aleutians.

Opposite, top. Baffin Island. *Bottom.* Canary Islands.

Baffin

A land at the top of the world, Baffin makes up one-third of Canada's Northwest Territories, but contains only one-fifth of the population. It is three times the size of Texas. It is a refuge filled with walrus, narwhal (the unicorn whale), beluga whales, polar bears, seals, Arctic wolf, caribou, fox, musk-ox, lemmings, and a wide sky of northern birds, such as ptarmigan, snow geese, and snowy owls. Each spring the rivers are alive with Arctic char surging down to the open sea, and some eighty different bird species migrate back to Baffin, including the rare peregrine falcon and ivory gull. The sea ice opens and polar bears head out to the floe edge to hunt seal and walrus. During the brief summer the land is carpeted with the most delicate of high Arctic flowers in a dazzling riot of color. In the fall hunters and fishermen head back to the communities, the tundra takes on its vivid autumn colors, and the birds head south. In winter curtains of northern lights dance eerily above the landscape, and travel is by dogsled. Here Inukshyuk Eskimos, rugged mountains jutting through massive ice-fields, the midnight sun, fjords, and glaciers stand as a testament to time. Baffin is a place to explore nineteenth-century whaling stations, a lost world of wooden ships, iron men, and mighty bowhead whales. Human history dates back 5,000 years here, and burial sites, kayak stands, meat caches, and ruins of winter and summer dwellings are part of the legacy of the ancient Thule Inuit.

Bali

In 1598, crew members on the first Dutch expedition jumped ship and went native. Today, Bali is just as seductive. Travelling through emerald-tinted carpets of rice fields graced with a thousand temples, it is easy to see why rice is recognized as a gift from the gods. This amazing terraced landscape is the result of fifty generations of farmers, or 1,000 years of practice. Although relatively small, the island is densely populated and culturally and economically one of the most important of Indonesia's 13,000 islands. Just off the eastern end of Jakarta in Indonesia, it is largely mountainous, with active volcanoes, and it has a great fertile plain to the south. The people are known for their artistic skill (especially wood carving), their physical beauty, and their high level of culture. Bali was converted to Hinduism in the seventh century, and was under Javanese rule from the tenth to the late fifteenth century. It is now an island of animist-Hinduism in a wide sea of Islam. The island harbors one of the finest one-day river rafting trips in the world, down the wild Ayung River, which flows through spectacular gorges and past magical stepped hills of paddies, from Ubud to the beach at Kuta.

Canary Islands

This group of seven islands off Spanish Sahara in the Atlantic Ocean constitutes two provinces of Spain, Santa Cruz de Tenerife and Las Palmas. The islands, of volcanic origin, are rugged. Mt. Teide (12,162 feet), on Tenerife, is the highest point. Pliny mentions an expedition to the Canaries circa 40 B.C., and they may have been the Fortunate Islands of later classical writers. They were occasionally visited by Arabs and by European travellers in the Middle Ages. The treaty of Alcacovas (1479) between Portugal and Spain recognized Spanish sovereignty over the Canaries; conquest of the Guanches, the indigenous inhabitants of the islands, was completed in 1496. The islands became an important base for voyages to the Americas, and were frequently raided by pirates and privateers. Wine was the main export of the Canaries until the grape blight of 1853; its place was taken by cochineal until aniline dyes came into general use; sugar cane then became the chief commercial crop. Today the leading exports are bananas, tomatoes, potatoes, and tobacco, which are grown wherever irrigation is possible. There is fishing on the open seas, and the Canaries, with their warm climate and fine beaches, have become a major tourist center, and a popular destination for hikers and divers.

Opposite, left. Truk.
Right. Faeroe Islands.
Bottom. Galapagos.

Devil's Island

The infamous Hell-On-Earth island is the smallest and southernmost of the Iles du Salut (Salvation Islands), in the Caribbean Sea off French Guiana. A penal colony founded in 1852, it was used primarily for political prisoners, among them Alfred Dreyfus, the island's first captive criminal. Although conditions were probably not as sordid as in other prison camps in French Guiana, the island's name became synonymous with the horrors of the system, and was featured in the book and film, *Papillon*. The penal colonies were phased out between 1938 and 1951. Devil's Island was originally uninhabited when the French first acquired Guiana in 1604. The wave-worn lava rocks and coconut palm covered beach, eight miles off the coast of Kourou, were first occupied by a handful of colonists who fled from the mainland to escape the dreaded yellow fever epidemic of 1765, hence the designation as a Salvation Island. Almost a century later, during the Napoleonic Wars, it became the infamous prison. Now the "Mansion de Dreyfus," the four-square-yard prison house, is empty, and the island only echoes with the cries of cruelties past.

Faeroe Islands

Belonging to Denmark, the Faeroes are a group of volcanic islands in the North Atlantic between Iceland and the Shetland Islands. There are eighteen main islands and four small, uninhabited ones. The Faeroes are high and rugged, heavily gouged and eroded by glaciers, and the rock makes a dramatic, often forbidding, backdrop to the scattered fishing and farming settlements. Around the fringes of the sparsely vegetated islands are the haunts of millions of seabirds. Their inhabitants depend mainly on fishing, whaling (mostly small pilot whales), and fowling. The earliest known inhabitants were Celtic. In the eighth century A.D. the islands were settled by Norsemen. In the early eleventh century they became part of the kingdom of Norway and were Christianized. The population was nearly wiped out by an outbreak of the Black Plague in the fourteenth century and was soon after replaced by Norwegian settlers. With Norway, the Faeroes passed under Danish rule in 1380, and they remained Danish after the Treaty of Kiel (1814) transferred Norway from the Danish to the Swedish crown. In World War II Great Britain established a protectorate over the islands after the German occupation of Denmark. In 1948 the Faeroes obtained home rule, and today they speak their own language, Faeroese, which springs from Old Norse.

Falkland Islands

Long overlooked, the windswept East and West Falklands, and some 700 smaller islands, lie 300 miles east of South America and 1,000 miles north of Antarctica. The islands boast a stark beauty: vast, rolling moorlands dotted by lakes and ponds; sheer cliffs plunging to the sea; velvety, surf-brushed beaches; and "stone rivers" carving parallel lines down a hillside. Clumps of tussock grass, nearly ten feet tall, provide an important habitat for elephant seals, southern sea lions, black-browed albatross, and five species of penguins. It is this abundant and remarkably tame wildlife that has lent the Falklands its unofficial banner: "Where Nature is still in charge." The islands are believed to have been discovered in 1592 by the English navigator John Davis, and are currently maintained as a British colony, though Argentina continues to claim the islands as their own, and calls them Islas Malvinas. South Georgia, a whaling settlement 800 miles to the southeast, and the South Sandwich Islands farther south—a group of small volcanic islands—are Falkland dependencies.

Opposite, top. Iceland.
Bottom. Komodo Island.

Galápagos

Charles Darwin, the English naturalist, made these volcanic islands famous after his historic visit in 1835. The animal life he found provided part of the inspiration for his theory of evolution, expounded in *The Origin of Species* (1859). Because of the unique species of plant and animal life—including the almost extinct giant tortoise—ninety percent of the land area is now a national park and wildlife sanctuary where naturalists can study living species arrested at various evolutionary stages. The Galápagos straddle the equator, and are 680 miles west of mother country Ecuador. They were discovered by the Spanish navigator Tomas de Bertanga, and named for the huge tortoises, in 1535. They were annexed by Ecuador in 1832. The capital is San Cristobal Island, also called Chathan Island. Four other islands have permanent human populations.

Iceland

Contrary to name implication, the island is not covered with ice, but in fact is gloriously green. Lush meadows, wildflower fields, and miles of rich tundra cover a landscape of remarkable variety. There are deep lakes, bubbling hot springs, powerful waterfalls, and snow-capped peaks. Iceland contains more lava than any place on earth, the largest glacier in Europe (the Vatnajökull), and myriad spectacular offshore skerries on whose cliffs nest a great diversity of seabirds, most colorful of which are the puffins. The human story of Iceland goes back over 11,000 years, and yet nearly every member of the present-day population, which numbers only about a quarter million, relates intimately to the history. Icelanders do so because their language has remained virtually unchanged in the millennium, and because their heritage is told in a treasury of fascinating sagas; legends of real-life heroes, trolls, witches, shape-shifters, and all manner of supernatural beings. Geologically the island fence-sits half in Europe and half in North America. As the two continental plates drift apart, magma wells to the surface, often quietly, sometimes violently, as it did January 22, 1973, on Heimaey Island. Today the optimistic islanders have come to terms with their volcanic home. They heat their homes, produce electricity, and bake their bread with volcanic steam. They even boast they have the only village church in the world warmed directly by the fires of hell.

Komodo Island

In the lesser Sundas group of the Indonesian islands west of Flores, this is the home of the giant monitor lizard called the Komodo dragon, the world's largest lizard. The Komodo dragon is the sole survivor of the carnivorous dinosaurs of 130 million years ago. Reaching up to ten feet in length, weighing up to 300 pounds, and living as long as 150 years, with 1½-foot-long forked yellow tongues, the lizards were perhaps the inspiration behind Chinese legends of fire-breathing dragons. The first Dutch expedition to the island was in 1910; two of the dragons were shot and their skins brought back to Java, resulting in the first published description. The island itself is hilly, desolate, sandwiched between Flores and Sumbawa, and surrounded by some of the most tempestuous waters in Indonesia, fraught with riptides and whirlpools. It is about nineteen miles long on its north-south axis, and about ten miles wide. Its highest point is 2,410 feet above sea level, and on the hill's flanks are skinny lontar palms and scruffy undergrowth. From the sea the island looks a far more fitting place for a prehistoric lizard than for the few hundred people who live in its lone village, Kampung Komodo.

Above. The Marquesas.
Right. Bali.

Opposite, top. South Island.
Bottom left. Vanuatu.
Bottom right. Mauritius.

Maldives

More than 1,200 coconut-covered atolls scattered across the Indian Ocean make up the nation of the Maldives. Shallow coral surrounds every island, creating a reef fantasia of color and texture, and beyond there are deep blue channels. Arabic dhows slice across the waters, their triangular sails arched in the wind. The islands were settled by Asians, who converted to Islam in the twelfth century and were influenced by Europeans from the sixteenth century on. Only 202 of the islands are inhabited in the 600-mile-long band that bends southward as if a god had cast emeralds at the equator. There are no hills in this country. The highest point of land is just seven feet above sea level, and from the shores of one island the others look like clusters of palm trees on the horizon. With the exception of turtles, iguanas, crabs, and birds, there is little fauna in the Maldives. It is in the crystalline and warm water that the islands come alive.

Marquesas

Jutting almost vertically from the ocean floor, the emerald-green Marquesas Islands form the most spectacular and remote archipelago in French Polynesia. Situated 875 miles northeast of Tahiti, the six major islands were settled over 2,000 years ago by Polynesian mariners from Samoa or Tonga. The first European to land here was Spanish explorer Alvaro de Mendana, who named the islands Las Marquesas de Mendoza, after his patron, the Viceroy of Peru. Mendana assumed he had discovered these islands en route to establishing a new Jerusalem in the Solomon Islands. The islands, because of their proximity to the doldrums of the equator, have always been a backwater of the Pacific. Volcanic in origin and geologically young, they rise like spires from the sea. There are no coastal plains, no reefs, and the valleys are deep, trenchlike, and lush. The first missionaries arrived in 1798, and in the following half-century different sects competed for the souls of the islanders. During this period of intense missionary activity Herman Melville jumped ship from a whaler and eventually wrote *Typee,* based on his experiences. The French painter Paul Gauguin was buried on Hiva Oa island in 1903.

Mauritius

Five hundred miles east of Madagascar in the Indian Ocean lies a tall-peaked volcanic jewel, a paradise of palm-dotted beaches, casuarina groves, and sheltered lagoons. Indians make up two-thirds of the population, followed by Creoles, Chinese, and Franco-Mauritians. English is the national language. Only thirty-six miles long and twenty-four miles wide, it is surrounded by one of the largest unbroken coral reefs in the world. All roads lead to Port-Louis, the capital city on the northwest coast. Visitors are greeted by ornate Hindu temples, Muslim mosques, and elegant colonial mansions. In the heart of the city is the Natural History Museum, where a replica of the extinct dodo (once in abundance here) is displayed. Nearby is the Bazaar Central, the shopping center of the city, a Sino-Indian marketplace where Indians peddle curry, saffron, and dried fish beside Chinese selling medicinal herbs and teas. Farther to the north is the fifty-seven-acre Pamplemousses Gardens. This is one of the world's finest botanical gardens, famed for its collection of palms and giant water lilies. The government of Mauritius carefully controls tourism growth to maintain the island's relative pristine environment, and to preserve the quality of the visiting experience.

Palawan

Narrow and sword-shaped, Palawan is the fifth-largest of the Philippine Islands, north of Borneo and between the Sulu Archipelago and the South China Sea. This little-visited island is host to the world's longest major underground river, running through the St. Paul Subterranean River National Park, one of the most amazing natural wonders of the world. This is a huge complex of some 200 caves, of which only twenty-nine have been explored by archaeologists since their discovery in 1962. Here were found human fossils carbon dated to 22,000 B.C., the oldest traces so far of *Homo sapiens* in the Philippines. With the "Tabon Man" were found Stone Age implements and relics of later eras, such as burial jars and kitchen utensils. Overlooking a bay studded with small islands, the entrance to the Tabon Caves is situated 100 feet above sea level, on a promontory facing the South China Sea. The large mouth leads to an equally imposing dome-shaped chamber, filled with stalagmites rising from the water. Bats are in abundance. At this writing no one has yet rafted this underground river, and it promises to be an extraordinary adventure.

Pitcairn Islands

On April 28, 1789, Captain Bligh and eighteen of his faithful officers were forced adrift in a twenty-three-foot boat by Fletcher Christian and other *Bounty* crew members. Then the *Bounty* was torched by the mutineers, who, seeking a safe haven from the reach of the British Royal Navy, took to this group of four small, precipitous, volcanic islands about halfway between New Zealand and Panama. Today the islands are administered by the British High Commission in New Zealand. They were discovered in 1767 by Philip Carteret, a British admiral, but were named after Robert Pitcairn, the midshipman who first sighted the main island. It was colonized in 1790 by those left from the H.M.S. *Bounty* and their Tahitian women. Their descendants still inhabit the main island, mostly in Adamstown, the only real village. In 1957 the remains of the *Bounty* were discovered here.

Robinson Crusoe's Island (Juan Fernández)

A stark, mountainous mote, ten miles long and about four miles broad, and some 360 miles west of Chile. It was this island upon which Scottish sailor Alexander Selkirk asked to be stranded in 1704 because of the mistreatment he was receiving from Captain Harding, commander of the S.S. *Cinque Ports.* His five-year castaway inspired Daniel Defoe's classic novel, *Life and Strange Surprizing Adventures of Robinson Crusoe of York, Mariner.* Though Selkirk was picked up by a British vessel and brought back to "civilization," evidence of his gripping adventure tale remains, such as the cave where he lived, and the lookout rock where he patiently awaited rescue. About 1,000 people now inhabit the island, the majority of whom make their living fishing for the large and abundant lobsters of the area. The entire island has been designated a national park. By the way, Selkirk's solitary reign ended in February 1709 when a couple of English privateers, one carrying his former captain, saw a distant, mysterious light, and sailed in to investigate. . . .

Saba

A beachless, five-square-mile extinct volcano, Saba has been variously described as a giant green gumdrop, an inverted ice cream cone, and "Napoleon's hat." It was used as "Cannibal Island" in the original *King Kong.* One of the northwestern Leeward Islands in the Caribbean, it is actually the cone of an extinct volcano rising to 2,854 feet. The peak is called Mount Scenery, and is the tallest peak in the Caribbean. The trek to the top is literally via staircase, one with 1,064 steps (I counted). The mountain is topped with mahogany and silk-cotton trees and enormous philodendrons. It is a terribly scenic island, known for its wall diving, but there are no sheltered harbors, and landing by boat is difficult. The chief settlement, called The Bottom, is in the crater of the volcano. Fishing and boatbuilding are the principal occupations.

Socotra

On his second voyage, Sindbad the Sailor landed on a desert island covered with strange trees, and a giant bird with "a leg as big as the trunk of a tree," the legendary roc. The island was Socotra, at the mouth of the Gulf of Aden, 150 miles northeast of Somalia. It means "Isle of Bliss" in Sanskrit, and was perhaps the legendary Land of Punt. The island was occupied by the East India Company in 1834, and became a British protectorate in 1866. In 1967 it chose to join the new, independent state of South Yemen, and now it belongs to the reunified Republic of Yemen. The mountainous interior rises to 5,000 feet, and is mostly barren. Its coastal plains and valleys produce dates, myrrh, frankincense, and aloes. Its exports include pearls and ghee. There is no air service to Socotra; the only way to visit is by Arab dhow, unless you are in the Soviet military, as there is a Russian base. The roc? It was probably related to the Mauritian dodo, long extinct.

South Island, New Zealand

Viewed from the air vast areas of this magnificent island appear as crumpled as a discarded piece of paper, with range upon range of titanic pressure-folded ridges stretching as far as the eye can see. The Tasman Mountains, the Richmond Range, the Saint Arnaud and Spenser mountains fan out northward and spill into stunningly beautiful fjords. Forested spurs soar to form the Southern Alps, in the folds of which are splendid glaciers and mirror lakes. In this massif there are more than 130 peaks that rise over 6,600 feet, chief of which is Aorangi, the Cloud Piercer, officially and prosaically named Mount Cook, after the great eighteenth-century navigator. On the coastlines the South Island is engaged in the ancient battle between ocean and land. In places the shoreline runs across the trend of the mountain ranges, to form a rugged shore deeply indented with bays and coves and landlocked havens, or deeply gashed with fjords.

South Shetland Islands, Antarctica

These include Elephant and Deception islands. The former is where, in 1916, Ernest Shackleton left twenty-two men and set out on an epic rescue attempt in a tiny open boat, hoping to sail 800 miles over the world's roughest seas to the edge of civilization. One hundred and five days after he left the island he returned on a Chilean ship and found all his men alive. Other than Mount Erebus in the McMurdo Sound region, Deception Island remains Antarctica's only active volcano. Long ago fire met ice and Deception Island blew its top, creating an eight-mile-diameter sea-flooded crater. On the black-sand Bailey Head beach more than 100,000 chin-strap penguins make their home. It is possible to take a boat into the flooded caldera of the volcano, called Neptune's Bellows; there lie the remains of a 1912 Norwegian whaling station and an abandoned British Antarctic Survey base. Neptune's Window, a break in the surrounding cliffs, is where American sealer Nathaniel Palmer became the first man to see the Antarctic Peninsula. At Pendulum Cove thermal springs bubble up to the water's edge, warm enough for bathing.

Opposite, top. South Shetlands.
Bottom. Sulawesi.

Sulawesi

An Indonesian isle shaped like an orchid, Sulawesi (formerly Celebes) is almost wholly mountainous, with many active volcanoes. Asian and Australian elements are commingled in the fauna, which includes the babirusa (a bush pig), the small wild ox called anoa, the baboon, some rare species of parrot, and a large number of crocodiles. Valuable stands of timber cover much of the island. The Portuguese first visited the island in 1512. The Netherlands expelled the Portuguese in the 1600s and conquered the natives in the Makasar War. It now holds four provinces in the republic of Indonesia. Tucked away amid the jagged peaks and fertile plateaus of south-central Sulawesi is the mysterious place called Torajaland. The Torajans are best known for their death cult—their elaborate, colorful feasts for the long-deceased during which animals are sacrificed in a bloody orgy designed to ensure that souls pass to the afterworld. The architecture of Torajaland is also celebrated. The roofs of the spirit houses rise at both ends like the ends of a ship. The houses symbolize the universe. The roof represents the heavens, and it is always oriented northeast to southwest, the directions of the two ancestral realms, according to Torajan cosmology. Some theorize that the cryptic Torajans are descendants of ancient visitors from outer space, and that their unique architecture represents the shape of their space ships. The name *Toraja* means "the people from above," and according to legend, their ancestors descended from the Pleiades, a star group in the constellation Taurus. Whatever their origins, the people, their art and religion, and the island are intriguing.

Tristan da Cunha

One of the loneliest spots on earth, this forty-square-mile mote lies halfway between South Africa and South America. It is formed almost entirely by a large (6,760 feet) volcano, which unexpectedly erupted in 1961. The population was evacuated to Britain, but returned in 1963. The island was discovered by the Portuguese navigator Tristao da Dunhain in 1506. In 1810 an American, Jonathan Lambert, landed here and proclaimed himself king of the island, which he named The Island of Refreshment. Two years later he disappeared on a fishing trip, and in 1816 the island was annexed by Great Britain. In 1938 it became a dependency of the colony of St. Helena. Nearby is an island called Inaccessible, the home of the flightless rail, an almost extinct bird.

Truk

Situated in the Federated States of Micronesia, Truk is a collection of tiny volcanic islands scattered across a vast expanse of ocean. Its total landmass is less than fifty square miles. Moen, the main island, rises above the clear waters of the world's largest lagoon, which during World War II sheltered the fleet of the Japanese Imperial Navy. The assault on the fleet by U.S. Naval forces in February 1944 (called Operation Hailstone) destroyed a significant number of warships, and was a turning point in the war. Now the lagoon is the site of the world's finest wreck dives. More than sixty submerged vessels and several downed aircraft rest in the depths of the lagoon, forming the world's largest underwater museum. In 1971 the legislature established the entire area as the Truk State Monument.

Below. Socotra.

Vanuatu

Formerly the New Hebrides Island, the country is, since 1980, the South Pacific's newest independent state. Although these eighty islands are traditionally and culturally a part of Melanesia, from 1906 they had been jointly ruled by a politically ambiguous compromise between the British and French. The republic comprises about eighty islands and inlets. The main islands, some 800 miles west of Fiji, are ringed with coral reefs. Five of them have active volcanoes, and earth tremors are frequent. Birds abound in lush forests. The indigenous people are Melanesians, a dark-skinned race, and those on the island of Pentecôst practice an incredible ritual known as the Land Dive. Volunteers stand atop a giant eighty-foot-high jungle tower and jump head first towards the ground. Inches from impact the fall is halted by two thin lianas tied around the ankles, and the jumpers bounce up and away from certain death. According to legend the land dive originated when a wife leapt from a banyan tree to escape her abusive husband. Though the lianas saved her life, he fell to his death. Today the ritual serves to ensure a prosperous yam crop, as well as allowing divers to demonstrate courage. Performed each spring by those who dare, these were almost certainly the world's first bungee-cord jumpers.

Zanzibar

A spice-laden coral island, Zanzibar floats in the Indian Ocean about thirty miles north of Dar es Salaam. Its first history is as a Persian sultanate, but by the sixteenth century it became a Portuguese trade depot. By 1700 Arabs had made Zanzibar a thriving center of East African ivory, cloves, and slave trades. It was made a British protectorate in 1890. Now it is part of the republic of Tanzania. The port of Zanzibar, on the west coast of the island, is the main town, a picturesque, cosmopolitan city with winding streets, colorful bazaars, and distinctive mosques, palaces, and cathedrals. In the nineteenth century the port was used as a base by European explorers and missionaries (including Richard Burton, John Speke, and David Livingstone) heading for the heart of the dark continent, the African interior. Henry Morton Stanley also departed from Zanzibar on his epic journeys to find David Livingstone, and to make the first crossing of the continent, which he accomplished in 999 days.